FALLING INTO PLACE

FALLING INTO PLACE

A Story of Love, Poland, and the Making of a Travel Writer

THOMAS SWICK

Foreword by Pico Iyer

ROWMAN & LITTLEFIELD
Lanham • Boulder • New York • London

Published by Rowman & Littlefield
An imprint of The Rowman & Littlefield Publishing Group, Inc.
4501 Forbes Boulevard, Suite 200, Lanham, Maryland 20706
www.rowman.com

86-90 Paul Street, London EC2A 4NE

British Library Cataloguing in Publication Information Available

Library of Congress Cataloging-in-Publication Data available
ISBN 978-1-5381-8177-5 (cloth : alk. paper) | ISBN 978-1-5381-8178-2 (ebook)

♾™ The paper used in this publication meets the minimum requirements of American National Standard for Information Sciences—Permanence of Paper for Printed Library Materials, ANSI/ NISO Z39.48-1992.

To Hania

And there is this about love: that its memory is not enough: for the soul retracts if it does not go on loving, whereas to have travelled once, however long ago—provided it was real and not bogus travel—is enough.
Freya Stark, *Perseus in the Wind*

Contents

Foreword: A Catcher of the Wry xi

Chapter 1: Love in Trenton 1

Chapter 2: Warsaw Lost .27

Chapter 3: Greek Tragedy 51

Chapter 4: Euro-Jersey Limbo 83

Chapter 5: Warsaw Found 107

Chapter 6: Philly Days 179

Chapter 7: Undivine Providence 217

Acknowledgments 247

FOREWORD

A Catcher of the Wry

Pico Iyer

THE CLASSIC BRITISH TRAVEL BOOK SERVES UP A DRY, INFORMED, OFTEN hilarious account of a well-educated traveler roaming across a foreign land. What contemporary American travelers regularly add to that is an openness to emotion and even urgency that suggests that something more than an exotic holiday may be at stake. In *Falling into Place*, a narrative as layered as its many-faceted title, Thomas Swick offers a singular blending of both forms, so as to take us around Europe (and particularly Poland) during Cold War days, while always keeping at the center of his attention the woman whose heart he longs to win: Hania.

Almost as soon as you enter this heartfelt and carefully wrought tale, you meet a kind editor who "looked like a young Virginia Woolf in tortoiseshell glasses," along with a soon-to-be-famous writer exuding a "drowsy gravitas." Clearly, our narrator is literate and graced with a wry sense of humor. He delights in language—"linguo-dexterity," as he might have it—and the land he's offering us a guidebook to is the one known as Observation. As soon as he travels out into the larger world, you also register how even the drabbest socialist landscape can be lit up by affection. The exteriors may be bleak and wintry, but his interiors are warm, rich with possibility.

A writer is only as good as his ability to read—not just people and himself, but also books—and Tom Swick is one of the best-read of modern American travelers. Not simply because he devours Waugh and Nabokov, but because he has steeped himself in the lineage of Cyril

Connolly and Patrick Leigh Fermor. One scene after another comes to vibrant life through the precision and grace of his prose: the sidewalks of Montreal "covered in grainy black-slush bootprints"; in Budapest, "bulbous sausages sizzling in the roseate glow of heat lamps." Beneath the descriptions is an elegant economy that can distill paragraphs into a single short sentence: "There are no comrades in a land of kissed hands."

Deeper even than that, by immersing himself in Poland's days of turbulence, the writer realizes that he is coming to understand America. He's seeing how resilient and full of lowercase solidarity people can be when life is most oppressive. And even as there's a smile in every corner of his often rich and complex sentences, his taste always runs to "quiet, unportentous beauty."

At some level, this is a story of loneliness and youth and ambition, a bildungsroman that just happens to be true. One minute you're experiencing the candlelit power of an Orthodox Easter in Greece; the next, you're reading of how both the writer's watches have stopped as he bounces through time-suspended Bulgaria. The first mandate of a book about travel is that it educates as much as it entertains; Tom Swick, writing in the wake of masters, pulls this off with unobtrusive ease and makes us feel, by the time we've reached the last page, that we've enjoyed an intimate glimpse into not just essential recent history but also into the sweet triumphs, and travails, of innocent aspiration.

Pico Iyer

Santa Barbara, California

January 2023

CHAPTER 1

Love in Trenton

In the fall of 1976, I returned home to New Jersey after a year in France. My dream was to become a travel writer, and in pursuit of it I had studied French in Aix-en-Provence and worked on a farm in Kutzenhausen, Alsace. Now I needed a byline, preferably a steady one. Making the rounds of newspaper offices, I stopped one day at the two-story brick building of the *Trenton Times*. I was not allowed to see anyone; this was the state capital's leading newspaper; I was simply handed a job application. There seemed little reason to play it straight.

What was your last employment?

Working on a farm.

What were your duties?

Picking cherries, baling hay, milking cows.

Why did you leave your last employment?

I got tired of stepping in cow shit.

May we contact your last employer?

Sure, if you speak Alsatian.

A few days later I got a call from the features editor asking me to come in for an interview. My reward for being original, and knowing my audience, or at least guessing it correctly. The previous week the editor of a small paper in Bucks County had emerged from his office and admonished me for wasting his time. "If you want to work in journalism," he had advised sorely, "don't spend your days picking cherries."

I drove the river road south from Phillipsburg, where I was living with my parents, back to Trenton. The features editor, Sally Lane, looked like a young Virginia Woolf in tortoiseshell glasses. She told me that the paper was owned by the *Washington Post* and that one of her writers, a young man named Blaine Harden, was exceptionally talented. The gist of the interview was that the editor—who, I later learned, had posted my job application on a wall in the newsroom—rarely hired people with no experience. But they would give me a three-month trial writing feature stories.

This worked out fine, for it allowed me to conceal the fact that I still wrote in longhand. I was possibly the last American journalist to do so. I knew how to type, but the typewriter was not a friend to the indecisive; it was good for deletions—a quick, brash row of superimposed x's—but, for additions, you had to take the paper from the carriage and scribble with your pencil between implacable lines and onto virgin margins.

In the evening, at home in Phillipsburg, I would write my stories, and then in the morning I'd get in my mother's car and drive the river road through Milford and Frenchtown (whose bridges I had worked on during summers in college), Stockton and Lambertville, the docile Delaware often visible through the leafless trees. The scenery was not as dramatic as in Provence, and the towns were not as picturesque as in Alsace, but there was a quiet, unportentous beauty to the place that suited my temperament, no doubt because it had helped shape it.

Once in the newsroom I'd type from my half-hidden, handwritten pages.

After I was hired full-time I bought my first car—a sea-green Datsun—and rented a studio apartment in Trenton. Most of the people at the paper lived in the more attractive surrounding towns, like Yardley, Lawrenceville, Princeton. Daisy Fitch, a fellow feature writer, had grown up next door to Albert Einstein. (In three years, her husband would receive the Nobel Prize in Physics, nicely continuing her relationships with scientific genius.) She was one of a dwindling minority at the paper, as it was increasingly being written by out-of-staters who swooped in for a spell and then left for careers at the *Post* or someplace equally grand. Many were Ivy Leaguers—this was a few years after Woodward and Bernstein made journalism as sexy as it was ever going to get—and, like Daisy, they often had interesting backstories. One of the city reporters, an Englishwoman, had started out in Washington as a secretary for Nicholas von Hoffman in the days when it was considered prestigious to have a woman with a British accent answer your phone. Celestine Bohlen, a young reporter, was the daughter of Charles "Chip" Bohlen, who had served as US ambassador to the Soviet Union in the 1950s. Mark Jaffe, a former fencer at Columbia, was living with the daughter of Lyle Stuart, the publisher made rich and famous by *The Sensuous Woman*. David Maraniss, who exuded a kind of drowsy gravitas, and for whom everyone predicted glory, came from literate stock: His mother was an editor at the University of Wisconsin Press, and his father was the editor of the *Capital Times*. I was told that I had just missed the Mercer County careers of John Katzenbach, nascent crime novelist and son of the former US Attorney General, and his wife, Madeleine Blais, both of whose auras still flickered in the brick building on Perry Street.

It was astonishing to find this assembly of near and future luminaries in Trenton, a city I had associated mainly with Champale, whose brewery we used to pass on our family drives to the shore. Add the fact that almost everyone had previous newspaper experience, and you can understand if I say I felt a bit out of place. All I brought to the party was an irreverent, if widely read, job application.

Though we all shared a love of writing. Most of the reporters burned with Watergate dreams, but many of them read the *New Yorker*, which in those days was less political. Nixon was gone; the Vietnam War was over; the country had just celebrated its two-hundredth birthday. Climate change, not to mention the demise of newspapers, was nowhere in sight. We would discuss Kenneth Tynan's stylish profiles—of everyone from Tom Stoppard to Johnny Carson—and the less flashy though generally more admired stories of John McPhee, the woolly high priest of creative nonfiction who lived and worked just up the road in Princeton. McPhee's approach—lulling readers with layers of meticulously researched facts delivered in effectively unadorned sentences, as opposed to dazzling them with arcane words and brilliant apercus—adhered more closely to the journalistic ideal of authorial self-effacement and reinforced an idea I had long ago realized: I didn't want to be a journalist.

The professionally limiting vocabulary was a deterrent. I would sometimes drop uncommon words into stories, usually humor pieces. It was partly my attempt to show that I belonged in the same room as people from Harvard and Yale. Unlike them, I had shown little interest in books until my junior year in high school, which was too late to get me the grades and SAT scores that would usher me into an elite university. I now read avidly, feeling a constant need to catch up, making lists of new words on yellow legal pads. And, like any nouveau riche, I took every opportunity to show off my recent acquisitions. Sally, the Bloomsburyish features editor, questioned the appearance of "wamble" in one of my pieces and I defended it by citing its use in a story by S. J. Perelman. She gently suggested that not many Trentonians read S. J. Perelman. I removed it, though with a feeling of satisfaction that I, a graduate of Villanova, had been asked to tone down my prose by a graduate of Barnard. Blaine expressed disapproval when I used the word "ululation" in a humorous essay. He was a graduate of journalism school, where he had been taught the importance of being easily understood. I was a former English major and assumed that other people would be equally delighted to increase their vocabularies. Though we both cherished our copies of E. B. White's *Here Is New York*, which we had discovered in a secondhand bookstore downtown and which I thought should replace *The Elements of Style* on every editor's desk. Years later I would read White's claim that he failed as a reporter because he always found that what happened to him while covering a story was more interesting than the story itself, a solipsistic reaction I recognized, by then, as belonging to most travel writers.

But a bigger problem for me with journalism—surpassing simplicity and invisibility—was its ephemerality. News was fleeting and so, inevitably, were the stories about it. And I didn't want to spend my life writing things that appeared and died; I wanted to write words that lived on or at least had a chance to.

Writing features was, at this point in my career, the closest I could get to this ideal—and still make a living—and I felt extremely fortunate to have been given the opportunity, from the start. Just how lucky I had been was brought home to me one morning when, for reasons long forgotten, I was asked to cover the funeral of a teenage girl in the Italian neighborhood of Chambersburg. I don't remember if she was murdered or killed in an accident, but I do remember calling the story in to the newsroom from a phone booth, the sensation I had at that moment of doing real journalism, and the relief I felt knowing I would probably never have to do it again.

No one would want to read that story today, while there might be some takers for my meditation on Evel Knievel, whom I interviewed in a Philadelphia hotel, or my profile of Ray Birdwhistell, the father of the study of kinesics. (It was accompanied by photos of the University of Pennsylvania professor making the facial expressions common to people in different geographical regions.) At Princeton, I found an English professor who had written a paper on talking dogs in literature, and I interviewed the sculptor and former prizefighter Joe Brown. A professor of art, Brown had also been the varsity boxing coach until McPhee's father, a physician at the university, had the sport banned. My farm experience came in handy one evening on campus when I covered a lecture given in French by Eugène Ionesco. The great playwright, after hearing an obvious doctoral candidate spout theory for several minutes before finally posing his question, replied, in words that spoke for over-analyzed artists everywhere: "*C'est très bon ça.*" Trenton State College provided me with a sociology professor who had researched the linguistic and cultural history of the hoagie.

My life was not all academics and daredevils. After President Carter's embarrassing trip to Poland—he announced upon arrival that he had left the United States for good and hoped to get to know the Poles intimately—I wrote a story about simultaneous interpreters at the United Nations. They were bright, curious, necessarily knowledgeable about a wide range of subjects and, oftentimes, learning more languages to add to their arsenals. "If you have everything it takes to do this job," one man told me a bit ruefully, "then you waste it on this job." Editors at the

Washington Post's Style section showed interest in this story but eventually passed on it, giving me one of my first writerly disappointments. Like everyone else, I dreamed of one day being in the *Post*. I wrote about a harness racing driver, a tobacconist, the owner of a general store, and a state government worker who repeatedly won the competitions in the back of *New York* magazine. Though I didn't realize it at the time, writing feature stories was giving me the perfect training to be a travel writer, getting me out into the world—in this case, New Jersey and its environs—to meet some of the interesting people in it.

One day in the newsroom a reporter by the name of Bob Joffee stopped me and asked if I had read *The Great Railway Bazaar*. "It's by this American," he said, "who takes trains through Europe and Asia and makes fun of everybody he meets."

I confessed I hadn't read the book. I vaguely remembered seeing the review of it in the *New York Times Book Review* a couple years earlier. I was envious of someone who had traveled so widely—I had not been outside of the United States other than in Europe or west of the Mississippi—and quietly outraged at his reported unappreciativeness. In France, I had devoured Evelyn Waugh's *When the Going Was Good*—one of the few books in English I allowed myself that year, putting myself even further behind in my reading—but the persona of the supercilious traveler seemed to fit an Englishman better than an American. We hadn't earned the right to cynical mockery and wry world-weariness.

When I finally picked up the book, I devoured it as well. I loved how Paul Theroux ignored the sights and wrote about the people; how he focused on little things—like hairy brown sweaters in Istanbul—that professional travel writers left out of their accounts in their pursuit of the Important, which was also, often, the Already Known. He seemed very much under the influence of Waugh but brought an American bluntness and seemed equally incapable of writing a dull sentence.

That spring, as the new feature writer, I was enlisted to cover the seasonal opening of the Great Adventure theme park in central Jersey. Before heading out, I ran into one of the photographers.

"So," the young man scoffed, "you gonna write the same old Great Adventure story?"

I hadn't been at the paper long enough to discover the cyclical nature of certain feature stories—the back-of-the-book equivalent to the predictable nature of most news stories. But I could imagine their numbing sameness year after year. And I was determined not to feed it. All day at the park that photographer's taunt ran in a loop through my head and helped me to focus—like Theroux and Waugh before him, visiting the Spanish-American Exposition in Seville in 1929—on the small, unexpected details; I concentrated on aspects of an amusement park on opening day that previous writers might have overlooked. Then, back in the newsroom writing the story, I added my personal stamp. Problems with the log-flume ride produced a line about "flummoxed flumers," a perhaps overly fanciful but heretofore unidentified group.

In a room full of crack editors and wannabe authors I had received the best writing advice from a photographer who didn't know he was dispensing it. No matter where I am on assignment, I still hear his jaded, tacitly challenging voice.

With spring came half-court games in the Lawrenceville driveway of Mike Norman, a former reporter who was testing a career in TV. I would tell Sally that I was thinking of taking the rest of the afternoon off to go play basketball, and she would say, "I think that's an excellent idea, Tom." It reflected her laidback management style but also the potency of April in Trenton. On the court, despite my small size, I could finally feel equal, in some cases superior, to my colleagues; it was the one place where I benefitted from having wasted my youth shooting hoops.

Many of us had taken music lessons when we were kids, and some of us still had our instruments, which included a clarinet, a saxophone, a trombone, a couple of trumpets. I played the drum, singular, which someone dug up from somewhere, and the house where we played basketball had a piano for Mike's wife, Beth. We came up with a name—Jersey Jam—and a slogan—"Gotta be jam 'cause jelly don't shake like that"—and played Big Band numbers, like "In the Mood" and "Pennsylvania 6-5000." This was years before the cult of retro and long before the return of swing; dipping into the past, we were ahead of our time.

Summer brought the Sun & Fun section, the paper's annual nod to travel writing. I hated the name but loved the vehicle. I wrote a story about my last summer in college, when I had forsaken the bridges over the Delaware and flown to London, where I found a job selling Bath buns and scones in a food hall on Oxford Street. It was the longest, and most personal, thing I'd written for the paper, and when it appeared on Sunday I felt the thrill, which had dimmed a little in the preceding months, of seeing my byline, my name in a newspaper atop thousands of my words telling of a colorful episode in my life.

Summer also brought humidity, which was familiar, and an acrid smell, which wasn't. Mark described it as the odor you got when you lifted the lid off a boiling pot of potatoes. The steamy imagery seemed appropriate. Driving down Perry Street I would see brown arms and faces hanging out of second-floor windows, seeking relief. Other sufferers sat on stoops.

I was sent to Atlantic City to write a story for our Sunday magazine. Gambling had recently been legalized, and my story was to be about the "last summer" of the fading resort. (This was before "the last" became an unavoidable, and usually inaccurate, construct in nonfiction book titles.) I knew the place fairly well; our family vacations had always been spent in nearby Ocean City, and more recently I had taken trips there with friends from school.

One of them joined me on Saturday night. During my last two years of college, I had had an incurable crush on Molly—there was something about her pale, sad-eyed, Irish face and her aloof, Main Line manner— but she had never shown an interest in being anything other than friends. She was one year older, more sophisticated and better traveled than I. Now, however, I had some European experience and a job in a profession that toppled presidents—not that I cared to do any toppling—and that had provided me with a paid room in Atlantic City.

We sat on the hotel terrace till well past midnight. It was one of the grand, doomed palaces that lined the boardwalk like ghostly vessels that summer. (They would almost all be demolished, an architectural loss greater than that of New York's Penn Station.) Sitting there, staring out at the blackened sea, would have been the perfect moment to reflect on the passage of time, changing mores. We may have discussed those things, but if we did my heart wasn't in the conversation. All I could think about was getting into bed.

I eventually did, while Molly took the couch. Lingering on the terrace, which I had hoped was a prelude to romance, had simply been part of her stalling strategy.

I returned to Trenton, dejected. In the newsroom, I struggled with the story. The fact that I was now writing on a typewriter didn't help. It was partly my fault for having invited her. I should have been out Saturday night, poking around, looking for, if not people to talk to, telling sights, illuminating scenes. As a travel writer, I had many things to learn, one of which was that the best travel writers travel alone.

The other problem was the monumental nature of the story. It was one thing to write about opening day at Great Adventure, when nobody was expecting much; it was something else to describe the death of a great seaside resort. I forced my words to reflect the grandeur of the subject and wrote one overheated lede after another. I grasped at obvious symbols— rolling chairs, rolling dice—when simple, sharp-eyed observation was called for. I stuttered and gushed, overwhelmed by significance. Atlantic City had handed me two defeats: the first as a lover, the second as a writer.

I had more success writing in longhand in my studio apartment. On my way home from France I had stopped in London, where I had met a young woman from Poland. Hania was working as a chambermaid in my hotel, the Mitre in Paddington, before returning to the University of Warsaw. We had one date, meeting at Marble Arch and then wandering about the city, and at the end of it I wrote my parents' address inside the book I had decided to give her: V. S. Pritchett's *Foreign Faces*, an account of his travels through Eastern Europe. I had presciently purchased it a few hours before my first sighting of Hania, behind the hotel bar, the night she filled in for the bedridden barmaid.

Our letters, crossing not just the Atlantic Ocean but also the Iron Curtain, kept our tenuous connection alive. Hers carried beautiful stamps and an unfathomable and literary—though I didn't yet know it—return address: *ulica Mickiewicza*. She wrote of coming to the States for a year, and, as happy as this made me, I felt obliged to tell her that Trenton was not the most desirable place, unless you were an ambitious young journalist. She was an econometrics major, with a charming epistolary style in a language that was not her own.

Hania arrived toward the end of the summer. I picked her up at Kennedy Airport and drove her to New Hope, where I was cat-sitting for Mark and Sandy. (If this were a novel, an editor would have insisted I change the town's name.) Once we reached the Delaware River, I drove on the more sylvan Pennsylvania side, where the road often curved under canopies of trees. The groggy passenger said they reminded her of Poland. Even after a twelve-hour journey, she looked beautiful, with her intelligent brown eyes, her short brown hair pushed behind delicate ears, her small, sharp features softened by immaculate skin. Other than New Hope, I had no idea where we were going.

The next day, Saturday, I had an assignment, again at the shore, to cover a billiards competition. One of the immense pleasures of my job was entering an entirely unfamiliar milieu and talking to the inhabitants, finding out about their specialized world, discovering its ways, learning its slang. It often produced interesting stories and made me a weeklong bore on a variety of subjects. And so it was in Asbury Park. When I was finished I rejoined Hania, who had come along for the ride, and we went to a diner, a fine American experience. Returning from the ladies room, she said with a smile: "I'm not sure I want to eat in a restaurant where employees have to be reminded to wash their hands."

Our glasses of iced tea arrived. I put a packet of sugar in mine and, looking up, saw that Hania was wearing a disapproving look.

"What's the matter?" I asked.

"You're making a lot of noise."

"I'm just stirring the sugar."

"It's rude," she told me.

We took a walk on the beach, and sat in the sand, overlooking the ocean. The situation was close enough to the one in Atlantic City that I didn't dare make a move.

That evening in New Hope some friends from the paper dropped by unexpectedly and invited us to join them at the Swan Hotel in Lambertville. Hania seemed a little disappointed. A short while later we walked across the steel truss bridge and joined my friends at the bar of the Swan. Surely, alcohol could only help.

Back at the house, I invited Hania to join me in bed. She climbed in, but I was still reluctant to act. "Are you going to take off your underpants, Mr. Swick?" she asked smartly and, within seconds, the relationship that had been built on the written word dispensed with the spoken one.

In the morning I told the story of Molly to explain my timidity and Hania said that if she didn't want to sleep with me she had no business agreeing to spend the night with me in a hotel. I didn't know if this was a European, or an Eastern European, or a Polish, or simply a non-Molly way of thinking, but I liked it.

Monday I entered the newsroom a different man. I looked the same, except perhaps for a delirious grin, but my life had entered a new, wondrous, long-awaited phase. Over the years, I had had more infatuations than girlfriends, had spent as much time pining as dating. I had never been in love, never felt the way I had that morning driving the river road to Trenton.

It was hard to concentrate on anything else. My notes on billiards, which I had taken so assiduously, now seemed unimportant compared to the flowering drama of my heart. Everything paled in comparison to that. I don't know how I wrote the story, so far were my thoughts from the subject. I worried that the emotional intensity of first love would have a detrimental effect on my work, which I had been so conscientious about. But I was helpless to do anything about it.

The house in New Hope became our honeymoon cottage. We'd take evening strolls through the town whose storybook sheen our romantic idyll enhanced. Sometimes we'd join the tourists waiting for exotic ice creams on South Main Street and then lose them as we made the short climb up the hill and over the canal back to our "home." Hania told me of the day, visiting the Tatra mountains with a girlfriend, when she had felt a physical longing for me. It gave me an immediate fondness for the Tatras. One day I came home from work and she showed me the copy of *The Sensuous Woman* she'd found, part of her personal discovery of America.

It was a bit of a shock when we had to move to my studio apartment in Trenton. But young lovers are generally oblivious to their surroundings, and the situation was temporary. Through some people I'd written about, I had found Hania a job on a farm in Bucks County. We had been, at the time, pen pals, remember; our present status, while hoped for on my end, was never a given. Also, the idea of living together was intimidating to someone who had grown up in a conservative family that was only an hour's drive away.

The owner of the eighteenth-century farm was a man who, every Christmas Day, dressed as George Washington for the annual reenactment of the crossing of the Delaware. Working there would have been a good experience for an exchange student, especially one with an interest in American history, but Hania felt isolated and occasionally slighted. Her forefathers had been landed gentry—her family's old manor house was now a large orphanage—but this lineage apparently carried no weight among Bucks County's gentleman farmers (just as my cherry-picking skills failed to impress its newspaper editors). Two weeks into Hania's stay I came for dinner—the hamburgers were made, I was told proudly, from one of the farm's cows—and, as I was leaving, Hania pressed a letter into my hand. As soon as the fieldstone farmhouse disappeared, I pulled off to the side of the road and tore the envelope open. In rounded cursive, Hania professed her love for me and her desire to live together. I drove home in a thick, euphoric fog.

She moved in. We slept together in my single bed. I'd write at the desk a few feet away, while Hania read atop the covers. I gave her, to complement *The Sensuous Woman*, White and Thurber's *Is Sex Necessary?* She discovered Fair Isle sweaters and frozen foods (Stouffer's lasagna became our favorite); we went to the Polish meat market in Lawrenceville, where I was introduced to *kabanosy* (garlicky sausage in long, thin, mahogany sticks). She made endless cups of tea. She coined a nickname for me, with an optional "y" at the end that allowed it to double as an adjective, which she used to describe everything from shoes to automobiles (usually curvaceous classics). After years without girlfriends, I now had one who had created an entire category of things imbued with my essence, an essence that had heretofore gone unrecognized, not least by me. Sometimes on Saturday evenings we would go out to dinner, often choosing rustic restaurants with stone fireplaces and French menus. In one such place we ate our first snails as Hania reminisced about her days at Catholic boarding school, collecting their Slavic cousins for export to France.

Occasionally Sally would invite us over. She and her husband, Sam Graff, lived in a three-story house near Cadwalader Park that had been designed by Sam's grandfather in 1907. Sam had worked at the paper before taking a job at the *New York Daily News* and, like Sally, was a proud Trentonian. The first time I saw him was at a newspaper dinner Sally had arranged at a Chinese restaurant downtown, a place owned by a woman Sally would champion in various locations, in none of which I ever saw a menu; you'd sit down, and Miss Julie would send out course after course of whatever it was she thought you would like. Thanks to dinners at Miss Julie's, I eventually learned to use chopsticks, a common skill today but one acquired back then chiefly by diplomats, travel writers, and spies.

Sam walked into the restaurant, straight from the train, like an actor onto a stage he was about to command. Long hair spilled from under his cloth cap and lapped against the collar of his trench coat. I took an immediate dislike to him, as I tended to do with outgoing people, whose attention-getting qualities I found objectionable because they were missing in me. Also, in my defense, and experience, such extroverts often turned out to be egotists. But Sam's exuberance, I quickly learned, was for life and not for himself.

The living quarters were on the second floor. You'd climb the right-angled staircase, greeted by a framed cartoon of a woman in a red beret firing a rifle and saying, "NOW, MES PETITS . . . POUR LA FRANCE!" I knew the French but not the artist, who, of course, was Lichtenstein. Upstairs, two large, high-ceilinged rooms overlooked Bellevue Avenue. The one on the right was lined with bookshelves that climbed the walls and overflowed with novels, essays, history books, art books, photography books, architecture books, and opera LPs as well as a record player and a radio that was perennially tuned to the classical music station in Philadelphia. More volumes teetered in piles on the floor. Buffy, the smiling Samoyed, wandered in at seemingly regular intervals. The house had such a warm, cultured, cluttered feel that, invited for dinner, we would often be there to hear the opening notes of Fauré's *Pavane*, signaling the start of WFLN's "Sleepers Awake."

I had never been happier. College had been a joy, especially for a new convert to books who was majoring in English, but it had lacked in extracurricular activities. Now, just a few years removed from it, I had a job *and* a woman I loved; this seemed like the straight American male's recipe for bliss.

The one dark spot, apart from Hania's soon-to-expire visa, was the fact that she had a boyfriend, a veterinarian-to-be by the name of Adam. Having never met him, I found I could usually put him out of my mind, but Hania worried about the pain she was causing. He had suspected there was a reason behind her coming to the States; that, as a presidential interpreter might put it, she had wanted to get to know an American intimately. She found a job as an au pair in Princeton, work she had done previous summers in London, and, returning from her first day of work, she told me, in a voice already familiar with life's cruel ironies—what I would come to think of as an unmistakably Polish voice—that the baby's name was Adam.

She rode the bus, meeting Americans most Americans never meet, and quickly learned that anyone who heard her name, on the bus or off, would reproduce it as "Tania"—preconditioned by Patty Hearst's nom de guerre. In the newsroom, I continued to write my feature stories and humorous essays, a few of which, I learned from Sally, had a conceit. Some evenings I filled in for the theater critic, returning to the office to write my reviews and then, at home, climbing into the bed made warm by Hania while reworking sentences in my head. The next morning, I couldn't bear to read what I'd written on deadline late at night. Deadlines, like typewriters, were something I never got used to writing on. However, the very act gave me a sense of (mediocre) accomplishment and of being an integral part of the running of the paper in a way that writing weekly feature stories never did. By necessity, daily journalism genuflected at the altar of productivity; quality was an occasional—though annually awarded—accessory.

One evening a group of us from the paper went to hear John McPhee speak at the public library in Princeton. He talked primarily about the architecture of his stories, using a blackboard filled with a complex diagram. His method, which I found hard to follow, seemed more mathematical than literary, and I came away highly impressed and deeply discouraged. I had gotten into this game because of my love of words.

I was still making lists of them, still reading steadily. The apartment was slowly filling with books, which I pulled from the shelves of the

poorly lit bookstore downtown. Hania and I pressed beloved books on each other, as bookish lovers do. I gave her James Thurber and Robert Benchley—fearing that Perelman would be the linguistic equivalent of LSD—and she countered with George Mikes, whose *How to Be an Alien* hilariously confirmed my belief in the importance of the outsider. Less enjoyable was Irwin Shaw's *The Young Lions*, the reading of which was a true labor *for* love. Though it told me that Hania came from a country that was still traumatized by the war.

At the end of November we made a trip to Montreal. Hania needed to leave the country to renew her visa, my first taste of the Cold War bureaucracy that would govern our lives for the next five years. But we set off in my Datsun full of eager anticipation, and the first night in Vermont—snow already falling outside, French already emanating from the TV—produced an additional layer of contentment. We had taken our love on the road.

In Montreal, officials at the American consulate told Hania that they couldn't renew her visa unless she presented proof that she would be returning to Poland. She had gone by herself to make the request, thinking that a new American boyfriend would only worsen her chances. As she explained to me back at our pensione, there was nothing she could give them that would *guarantee* she wouldn't stay in the States. She bristled at the lack of logic.

Our stay in Montreal lasted longer than expected. Every night, snow fell, hushing the city, and, every morning, the inhabitants calmly went about their business, which, in the case of US foreign service officers, included the unintentional sabotaging of our happiness. Hania returned to the consulate and then to our pensione, each time wearier and more frustrated than the last. She explained the latest developments quickly, as she did everything, and sometimes incompletely, and, when I asked her to clarify, it only frustrated her more. Lying on the bed we had, not our first fight, but our first misunderstanding. Tension had finally entered our relationship. It was the first time I saw Hania cry.

I feared our time together was coming to an end and chafed at the fact that a government—my own!—and not a person would be to blame. You can't challenge a consulate to a fight.

Not that that was my style.

We made long-distance telephone calls to Poland and New Jersey. Sally's father was a lawyer, as was mine, but hers had better connections. We attended a concert by Gilles Vigneault—the Jacques Brel of Quebec—and heard him sing, along with thousands, "*Mon Pays*." "My country is not a country," they roared in French, "it's the winter." We went to a talk given by the novelist Mordecai Richler, who had become the American media's go-to man for thoughts on separatism, at the time a very hot topic (which meant, of course, that Americans got only the anglophone view). I'd ask people for directions in French, which Hania also spoke, and, when neither of us understood the response, she would ask again in English. One evening I went by myself to a working-class neighborhood and walked with crowds down sidewalks covered in grainy black-slush bootprints and lined with French-named bars and taverns. Even under duress, the travel writer in me was following the locals instead of the tourists.

Hania had an aunt and uncle living in the city, and one day we paid them a visit. Zbyszek stood on the street corner in his tall fur hat—having exchanged one cold country for another—and bent down to kiss his niece three times on the cheek. We walked to their apartment, where I answered questions in French between heavy rounds of Polish. At one point during the dinner I noticed that Hania and her aunt and uncle were eating with their forks in their left hands and their knives in their right. Mid-meal I made the switch—quite easy for a lefty—and said goodbye to my two-decades-old food delivery system.

A cousin eventually came through with a statement that Hania had to return to Warsaw in the spring or all her university work up to that point would be discounted. He was a professor but not *her* professor; the professor she had asked for help had refused her request. With a new, six-month visa in hand, we crossed the border into a country I now saw a bit differently.

But I still wanted to show Hania the best of it. On our drive south, I spent hours searching for the quintessential New England inn. I'd pull up to one, walk in, and then return to the car after a few minutes, explaining that it didn't quite meet my image of the ideal. We stopped at a handful of charming, and high-priced, also-rans until it got quite late and, with no more inns in sight, we found a room in a chain motel. The outing, I thought, had the makings of a Benchley story.

For the holidays we headed up the river road, passing through the bridge towns festooned with lights. My parents were happy to see us, of course, though my father had not warmed to Hania; she was too independent and forthright for an American male of his generation. She relished a good argument, which she saw, in the European tradition, as a lively discussion, and the two topics that were unwelcome at the American dinner table—religion and politics—were the two that Poles lived and breathed. At the same time, she adhered to Old World ways, allowing (teaching) me to open doors for her and, on visits home, using my father's kit to shine my shoes. My brother Jim walked in once while she was vigorously buffing a pair of my loafers and gave a look that suggested he couldn't believe my luck.

On Christmas Eve we drove across the river to Easton for the midnight service at Trinity Episcopal Church. It was where I had been baptized, confirmed, and recruited as an altar boy, often for this service, during which, in my teenage years, I fantasized about the girlfriend I would bring to it one day. Nancy Stableford, Roger Angell's sister, sat two pews in front of us; she had married a biology professor at Lafayette College and given the church a cherished—at least by me—literary connection. Father Gill climbed the steps to the pulpit, where he quoted G. K. Chesterton and Charles Schulz. I exulted in the music, the incense, the candleglow, the hint of E. B. White, the nearness of Hania, who, I now noticed, looked distraught.

It was, she reminded me afterward, her first Christmas away from home. What the officials at the American consulate in Montreal did not know—how could they?—was that Hania loved her home and, unlike many of her compatriots, had no desire to make a new one somewhere else. This fact kept me from questioning the authenticity of her feelings for me—she couldn't be using me to get a Green Card when she didn't want a Green Card—but it highlighted the challenge that awaited if I wished, as I did, to make her my wife.

Jersey Jam rang in the New Year at a party at Mike and Beth's house. During our set, Hania disappeared into the kitchen, where she read, she told me later, an informative calendar item about bees. She was not one to stand appreciatively during someone else's performance. Or perhaps our playing was so bad it drove her away. Sometime after midnight, Beth came up to me, while Hania sat on the couch with a young reporter, and described him as "putting the moves on your little Polish girlfriend." She was small, but I had no doubt she could look after herself.

I wrote a story about Montreal for our Sunday magazine—a kind of "Letter from Quebec"—that ran with a caricature I had done of Mordecai Richler. (Inspired by Thurber, I'd been drawing amateurishly for several years.) David Maraniss departed for the *Washington Post*, and Sally's cousin, the most swashbuckling of the newsroom characters, headed off to Africa. One of his letters home was read aloud to me by Mark, who noted the recurring theme—a steady diet of bananas—and how it was judiciously, not excessively, repeated for ideal effect. "Now *that's* comic writing," he said, implying, it seemed, that what I had been doing wasn't. Blaine, who was soon to join Maraniss at the *Post*, wrote a story about a Stockton quarryman that began with the sentence "Stash Gorsky busts rocks." I thought it a perfectly banal lede, but it's the only one from those days that has stayed with me, outliving even my own. I use it in travel writing classes now to show how words can be used to convey appropriate sounds. Dan Laskin, reporter and trumpeter, got a job at *Horizon* magazine in New York, which led to his becoming the first person of my acquaintance to eat sushi. This distinction, along with the new editorial position, seemed to suggest that, gastronomically and professionally, he had left us all behind. Though he still lived in the area, having moved with his girlfriend, Jane Cowles, a graduate student at Princeton, into the vacated Maraniss quarters that were part of the Jersey Jam duplex. Like many of the reporters, I was still eating a good number of my lunches in the newspaper cafeteria, whose cheese omelets I always followed with a KitKat, savoring each miniature girder so intensely—the way I had eaten Toblerone after student meals in France, keeping the bitten-off mass from reestablishing contact with my teeth—that the undressed wafers scraped the roof of my mouth. Also staying in-house were the investigative skills of one reporter who got his hands on a salary list and revealed it in protest of what he considered our unconscionably low wages. Not surprisingly, my name appeared at the bottom. "Yes, but you have to understand," one of the editors apparently explained to Celestine, "hiring Swick was like hiring a cowboy."

I admittedly have no head for business, but it would have seemed churlish to ask for more money to do what I was so happy doing.

Though I was beginning to see the repetitive nature of feature writing, as the same events from last year demanded coverage again. The beauty of travel writing, it seemed, was that you were always seeing new places, despite the fact that most newspaper travel sections returned again and again to a confoundingly limited number of seemingly approved destinations.

One evening early in our relationship Hania and I had stopped at the pancake house in Princeton and sat at a table by the front window. During a lull in the conversation I had looked out at the people walking along Nassau Street and thought, "This is the person I'm with now, to the exclusion of all others." For a shy, single young man, it was a daunting prospect—not because I was antisocial, but because I dreamed of being the opposite. Loneliness, if it's not depressive, comes with an active sense of possibility, the everlasting hope of the new. Companionship is the fulfillment of that hope and, if it's tied to commitment, it eliminates the thrill of anticipation. Life becomes fuller, richer, but also, in one important arena, decided, robbed of enticing potentialities. You've found your desired place, and you'll travel no further.

These were not my thoughts as Hania prepared to leave that spring. I knew I might never see her again. The fear lingered that, after a few weeks back in Warsaw, reunited with her boyfriend, she would view all this as a harmless interlude. (Once, when I told her I would always love her, she cautioned me not to say that, because one never knew what the future held.) We drove up the river road so she could say goodbye to my parents. We visited Sally and Sam, who were sympathetic and comforting, and Hania hugged Buffy while a Bach fugue played. The night before her departure, she couldn't locate her plane ticket; we scoured the apartment and finally found it sitting in the kitchen garbage basket. I told her, honestly, that I hadn't put it there.

In the morning we headed out early, hours before her plane was to depart, and drove through the Holland Tunnel to have lunch in Midtown. I found a parking space easily on a major cross street, and we walked around the corner to a gyro place I knew. When we got back, about half an hour later, the car was gone. Now reading the signs, I discovered that, in my impaired mental state, I had blithely parked in a no-parking zone. This was, in the crime-ridden New York of the '70s, very good news. We flagged down a policeman who pointed us in the direction of the tow-away garage, from which, after paying a high ransom, we retrieved my Datsun, with Hania's suitcase still in the trunk.

We made it to Kennedy on time and said our goodbyes in the Pan Am terminal. I stood and watched in despair until the blurred plane wrenched back from the gate. As it rolled inexorably toward the runway, I headed inconsolably down the corridor, the two of us now moving in opposite directions. When I got in the car, the sight of the empty passenger seat produced a new wave of tears.

I was once again emotionally racked, but my work didn't suffer this time because it was all I had. A small section of the newsroom was taken over by new machines, which sat in a row like deep-set televisions with keyboards. Departments took turns getting operating instructions. When my turn came, and I saw the letters I typed appear on the screen, I did not rejoice at my coming emancipation from the typewriter. I resented the cumbersome log-on process and the stubbornly blinking cursor, which seemed as hostile to slow, ruminative composition—"*Come on. Let's go. Hurry up. I'm waiting.*"—as the typewriter was to revision (neither of which newsrooms had much use for). It threw up another obstacle between the thought and its appearance on the page and seemed horribly unsuited for anything literary. I was such a traditionalist I could not embrace the thing that would make my writing life infinitely easier.

At home I started reading *Anna Karenina*. I wanted to learn about love, the emotion that had consumed me for the last ten months, and a Russian master seemed the ideal instructor. He had me, like so many others, at the first sentence. I wondered if it could have been written on a word processor.

Letters from Hania arrived, to my enormous relief, and I wrote back, sometimes in a comic-erotic style. The plan was that I would leave the paper and come join her in the fall. To hell with the *Post*. It was a purely personal decision, but it came with ancillary professional benefits. I would get to see a country hardly anyone, with the exception of V. S. Pritchett, had written about. My one frustration with my year in France was that most of it had been spent in Provence, which had been done: Lawrence Durrell, James Pope-Hennessy, M. F. K. Fisher, Cyril Connolly. (Of course, one decade later, this all-star lineup would not deter Peter Mayle.) Poland gleamed as an unknown, an enticing potentiality.

I enrolled in evening Polish classes at a local college. The teacher was a prim, soft-spoken woman, which was fortunate, because the language was a bitch. About the only word I had learned from Hania was *siusiu*, which is what she would go off into the woods to do on long car journeys after drinking lots of tea. Krystyna—the "tyn" pronounced as "tin," not "teen"—increased my vocabulary, mostly through greetings and polite expressions. Poles, I was learning, were a very punctilious people. ("What the hell does 'punctilious' mean?" one of the non-Ivy reporters once shouted at me in the newsroom while reading one of my humor pieces.) But the grammar was beyond teaching, or I should say learning, in a classroom. Polish, like Latin, is an inflected language, so words change depending on how they're used in a sentence. The word for dog is *pies*, pronounced like the French *pièce*. But if you say you have a dog it becomes *psa*. If you go for a walk with your dog it becomes *psem*. If you give your dog a bone it becomes *psu*. And so on. If you want to describe your dog, the adjectives change as well. God help you if you have two dogs. Ahead of me, clearly, was another labor for love.

Books about Poland were divided almost exclusively into two types: histories and political tomes. Both often came swaddled in bright-red dust jackets. What I wanted I couldn't find: something that would tell me about contemporary culture and everyday life. I craved a book that would give me some idea of what to expect, what I was getting myself into, and Cold War analyses weren't sufficient.

I asked Krystyna for the names of contemporary Polish writers and, without hesitation, she mentioned Marek Hłasko. Princeton's Firestone Library, in those days open to the public, had some of his work in translation. His style was spare, sort of the way Hemingway might have written if he had been working class in a socialist country and in possession of a (bleak) sense of humor. Of course, the political undertones were lost on me. In fact, I found myself, thanks to Leo, wandering over to where the universal Russians sat: Chekhov, Turgenev, and Vladimir Nabokov, the playful, erudite, Slavic lover of language(s) I had been looking for.

One day I boarded the train to New York—why take a car into the city?—to apply for a visa at the Polish consulate. It occupied a beautiful Beaux Arts mansion at the corner of East 37th Street and Madison Avenue. The entrance to the visa section was through a door to the right of the main entrance, and walking through it I thought of Hania in Montreal. Though it seemed to me that the dread experienced by the citizen of a capitalist country on a visit to the consulate of a socialist one is greater than when the situation is reversed, especially when the latter citizen is opposed to the system under which she lives. In any case, the place made me queasy, despite the high-society touches that, of course, were unbecoming of a people's republic. *These* people, barely visible behind glass, were our enemies, and I had come to ask them for a favor. A really big favor. My fate, without exaggeration, lay in their hands.

For some unknown reason, they allowed my life to follow the course I had besottedly chosen for it. Though only temporarily, through a tourist visa.

I gave my notice at the paper. Leaving a promising career in journalism to pursue a woman behind the Iron Curtain was, for some in the newsroom, the most remarkable thing I had done since filling out my job application. (Bob Joffee, on meeting me my first week, had expressed disappointment that I appeared so tame.) My days at the *Times* were bookended by radical acts.

After work one day I drove out to the covered bridge I had discovered on the way to one of my assignments. The banks of the stream were lush and green. I had never spent much time in nature and, since that summer in London, I had become a lover of cities. But Perry Street was not a place for contemplation. Nor, even, was Frenchtown. Out in the woods, birdsong graced my crossroads reverie.

My brother Jim rented a small truck to move my furniture, and my starter library, back to Phillipsburg. My mother came along to help with the cleaning. Just before we finished, the telephone rang. It was Hania, calling from Warsaw. We chatted briefly—overseas calls were expensive—and when I closed with "I love you" I was answered with silence. It was probably the connection, but the unechoed sentiment filled me with concern.

So much so that I barely spoke as I drove up the river road. My mother complained that I wasn't good company. This was not how I had imagined leaving Trenton, but my panic served a purpose: Freshly alarmed about the future, I gave no thought to all that I was abandoning.

CHAPTER 2

Warsaw Lost

I SAT ON THE STONE STEPS OF THE SIXTEENTH-CENTURY FERME DES Fleckenstein and opened a box containing xeroxes of my feature stories. Dany sat next to me, as tan, strapping, and addicted to Gauloises as I had remembered him. It was a soft summer evening in Alsace, and I was proudly showing the fruits of the job that the farm had helped procure for me. They constituted a modest, shuffled collection, but to me they were seeds from which magazine pieces and even books might grow. Dany glanced at them with an apathy that was tied to envy, not because he couldn't read them, but because they had resulted from a profession that, unlike his, produced lasting results.

Nothing had changed. Geraniums still bloomed in the window boxes of the half-timbered houses, and the Protestant and Catholic church steeples still towered over the village. Papa Fritz, who resembled Henry Fonda in *The Grapes of Wrath*, was still kindly and his wife still businesslike. Dinner-table conversations were still conducted in Alsatian, that musical tongue, with snippets of French thrown in for my benefit. (Whenever I found a dead fly at the bottom of my bowl, I was told the day's special was "*soupe à la mouche.*") The barn was still attached to the house, and the windows still held no screens. Lulu the beagle, chained to an ancient doghouse in the courtyard, still barked at the cows as they were led out to pasture. In my bedroom, a photograph of Albert Schweitzer still stood atop the dresser. And Dany was still dreaming up new schemes—this summer's was red currant wine—to relieve the monotony of farming's endless repetition and the boredom of living with his parents.

I had stopped in Kutzenhausen on my way to Warsaw. Alsace and Poland share a number of similarities—a love of pork, a slew of blondes—but the most notable is an unhealthy geographical position. Alsace is squeezed between France and Germany the same way that Poland is caught between Germany and Russia and—if you think about it—the same way New Jersey is stuck between Pennsylvania and New York, though those two entities, while also sizable, have been mercifully nonaggressive—if you discount the jokes emanating from Manhattan. Living in Poland, for someone who had grown up in New Jersey and worked in Alsace, seemed a logical progression, one that would solidify my affection for the underdog.

In one of his first letters to me after I left the farm, Dany had written that "*notre porte est toujours grande ouverte pour toi*," never realizing, I'm sure, how quickly I would be back on the Mall family doorstep. But, just as he had two years earlier, he appreciated the companionship as much as the help. I fell back quite easily into my farmhand persona, even though now I was more guest than worker.

One afternoon some relatives visited from Strasbourg with the news that a young Polish woman was staying in their house on a Methodist church exchange. A few days later I was driven down to meet Jolanta, who told me that her father, in addition to being a minister, was also the director of an English school in Warsaw that had been founded in 1921 by a group of American Methodists. She suggested that, after my arrival in the city, I go see him about a job. The continuing role played by an Alsatian farm in my professional life seemed slightly Dickensian.

Letters to and from Hania traveled a bit faster now that they remained in Europe. I made reservations—plural, as there were no direct trains between Strasbourg and Warsaw—and sent Hania the date and time of my scheduled arrival. She wrote back expressing concern about her mother, who was experiencing pain in her leg.

Then one evening, after my goodbyes to his parents, Dany drove me to Strasbourg, where I boarded a train to Cologne. The cathedral grew gradually in the window, replaced, very quickly, by the late-night confusion of a foreign station. I lugged my big leather suitcase, reminiscent of a steamer trunk, and hobbled past platforms, miraculously discovering the correct one a minute before it emptied of travelers.

We chugged eastward into the night. I was like Theroux, taking a train to some strange land, though I wasn't on a solo adventure; I was headed toward a reunion. Someone was waiting for me at the other end, the secret fact that, even more than class, distinguishes one type of passenger from another.

The thought gave me comfort, of course, especially when we entered Poland. Could I have conceivably slept through the Berlin Wall? It seems impossible, but my next memory is of the Polish border and the young male officers in their tight-fitting, olive green uniforms. They were my first glimpse of Hania's compatriots—my competition—and they had a swagger that went beyond the official and seemed almost sexual. Immediately I felt out of my depth.

A flat, dry land rolled past our window. A German woman looked out at a field of stunted wheat and said, "We're not in Kansas." We weren't in West Germany, either. The bright sunshine could not dispel a general air of fatigue.

Around midafternoon a tower rose in the distance that I knew was the Palace of Culture. Like the Cologne cathedral, it dominated the horizon, and grew larger and more distinct as the train moved closer. We came to a stop at what looked surprisingly like a provincial station, and I uncertainly grabbed my suitcase. Not everyone disembarked; the train was heading on to Moscow. Outside on the platform, I bumped against waves of ruddy-faced strangers and suddenly saw Hania, smiling wanly in a baggy brown dress.

We got a tram to her apartment, or rather her mother's, where she was now living. Her mother was in the hospital. Children played in the yard of a gray, four-story block. "The future of Poland," Hania said softly.

The apartment had a bedroom, which was rented out, a tiny kitchen, and a tiny balcony. The couch in the living room turned into a bed that abutted the dinner table. I helped Hania make it. Then I held her—the long-awaited embrace—but she was elsewhere. She was sorry, she said; her mother was dying.

She returned to the hospital, and I went out to buy some water, picking up a bottle of clear vinegar instead. *Wódka*, so similar to the word for water, *woda*, would have been a more understandable and useful mistake.

Back at the apartment, I turned on the television. The anchor reading the evening news gave me no hope of ever penetrating the thicket of sound that emanated from his mouth. I acquainted myself with the ingeniously engineered Polish bathroom, where no space was wasted, not even that above the tub, where drying laundry hung from half a dozen rigged lines. Squatting to shower with the hand-held nozzle, I sprayed the clothes as thoroughly as myself.

In the morning I was driven to the hospital on Banacha, a street in a neighborhood not far from the airport. A friend of Hania's mother greeted me outside her room, and, lost for words, I kissed her hand—having read that Polish men still did this. Then I was taken in to see the woman I had hoped would become my mother-in-law.

That night, sleeping on the sofa-bed in the living room, Hania and I were awakened by a noise. Getting up to inspect, she found a small painting had fallen off the wall.

At the hospital the next morning, Hania learned that her mother had died during the night. The fallen picture, she assured me, had been the portent.

Hania was an only child, and her father had been killed in a motorcycle accident when she was a teenager. This was years after he and Halina had divorced; the marriage, never ideal, had been irreparably damaged by Halina's imprisonment. In fact, she had given birth to Hania in prison, after being tried, while pregnant, for activities against the state. Her crime was providing assistance to a man sent back to Warsaw from the government-in-exile in London.

Despite the early traumas—once, on a visit to the prison, little Hania screamed at the sight of a woman she failed to recognize as her mother—she claimed to have had a happy childhood. As a teenager, she had made the decision to enroll in Szymanów, the all-girls Catholic boarding school her mother had attended, because she wanted an education free of socialist propaganda. (When Hania first told me this, I thought back, painfully, to my teenage concerns.) There, she had forged a number of the intense, lifelong friendships one makes at that age when living in close quarters with one's peers. Some of these young women lived in Warsaw, as did most of her childhood and university friends, along with a crowd of aunts and uncles and cousins, some real, some honorary, in that tight network that is all the stronger in countries where the number of true relatives has been reduced by war.

But now she seemed heartbreakingly alone. I was there to give support, of course, but I couldn't help her plan the funeral, or arrange the burial, or deal with paperwork; I couldn't even join in her grieving. I was in many ways an added concern, a helpless foreigner who was unable to buy groceries. Though I did serve a purpose: Hania said that the story of my vinegar purchase was the last thing that made her mother laugh.

Halina Matraś was buried in Powązki Cemetery, which is to Warsaw what Père Lachaise is to Paris, though with more of a heroic, martial character. I was assigned to a motherly friend, and, after the casket had been lowered into the ground, I saw Hania being escorted out by a band in black, including a young man who led her on his arm. I felt an immediate tinge of jealousy, until I told myself he was probably a cousin. But it was a shock, after the exclusive idyll of Trenton, seeing Hania in her world.

Adam the ex-boyfriend stopped by the apartment. He was a well-built, handsome man with a dark goatee. He pulled me into the bedroom—the boarder was away—and opened a Polish-English dictionary. After searching for a few seconds, he indicated the verb for "love" (*kochać*). Next, he moved to "P" and showed me "*pomagać*" (help). Then, looking me in the eye, he pointed in the direction of the living room, where Hania sat. I had just gotten my first lesson in Polish gallantry.

This was not how I had pictured our new life beginning, but in retrospect it was a fitting start, a visit to a hospital followed by one to a cemetery. I was given a sense of the shadows that hover at the edges of all life but seem somehow to intrude with more frequency into the Polish edition. I had read that the first business to open in Warsaw after the war had been a flower shop and thought it touching—this triumph of the romantic over the practical—until I realized that the flowers were needed to adorn all the fresh graves.

I went to see Jolanta's father. He lived in an apartment above his school, which occupied a prewar building on Plac Zbawiciela, a downtown circle bisected by trams. Time in Poland was divided into two periods: before the war (*przed wojną*) and the thwarted, disaffected decades since. Mr. Kuczma, a short, stolid man with combed back hair, seemed not at all discouraged by the news that I had never taught; the fact that I was a native English speaker, a rarity in Poland in those days, was sufficient. All I needed to do was change my visa, and he'd give me a job at his English Language College.

The visa office sat a short walking distance from the school, in an imposing building on ulica Krucza (Raven Street). I was surprised to find it crowded, mostly with Arab men and their Polish girlfriends. Hania, following their lead, helped me fill out my application. We were told the response would take a few days.

We got a train north to the seaside town of Sopot, where one of Hania's aunts lived. She needed a change of scenery; also, she wanted to show me her country. Yet even Sopot, with its salt air and Grand Hotel and late-season tourists, wore the same tired, threadbare look that Warsaw did. (The Grand Hotel, of course, was *przed wojną*). Trenton was not a beauty spot, but it had buildings and neighborhoods that lifted one's spirits; here, nothing man-made, apparently, had escaped the shoddy touch of socialism. It depressed me greatly though, because it was her home, Hania took it in stride, the same way I did the poor sections of Trenton that not only saddened but also frightened her.

Though she was shocked the day we took the bus into Gdansk and saw flags with swastikas flying from buildings. Turning a corner, we found a vintage car and young men in Nazi uniforms milling around cameras. They were taking a break from the filming of *The Tin Drum*.

A cousin, Iwona, and her husband, Zbyszek, joined us one day. Zbyszek, like most Polish men, was a versatile handyman—public disrepair had given birth to private initiative—and when he asked me what skills I possessed I told him I sometimes drew caricatures. He was not impressed. A few hours later he grilled me on my government's treatment of Native Americans. I was too taken aback to think to ask him about Polish anti-Semitism. But it was a strange introduction to a member of a family I hoped to join and about the only instance of anti-Americanism I experienced in two-and-a-half years in Poland. For the most part, moving from France—the university, not the farm—to Poland was like transforming from a pariah into an idol.

We returned to Warsaw and went straight to ulica Krucza, where, my stomach churning, my life in the balance, I picked up a work visa that was good until the first week of February. To celebrate, I walked down ulica Piękna (Beautiful Street) and entered the embassy of the United States. I badly needed to use the men's room.

I told Mr. Kuczma that I was available for teaching, and he gave me five classes, from beginners to advanced. Classes were made up primarily of high-school and college students—one had to be sixteen to enroll in the evening school—and teachers taught on alternate days. I got to practice on a class of another American teacher, my sole compatriot, a Mennonite woman from Iowa who had not yet returned from vacation. "Just walk in the door," Mr. Kuczma advised me, "and you'll be fine."

I did, and I wasn't. The lesson that day was on the conditional mood, and the drill moved back and forth among the three variations: "If it rains, I will . . ."; "If it rained, I would . . ."; and "If it had rained, I would have . . ." It's easy and obvious when you're sitting at a keyboard, but it gets confusing when you're standing in front of a roomful of students for the first time in your life, in a city you've just moved to in a country behind a curtain. Worsening my disorientation were the faces of the female students, who were in the majority and of an attractiveness one didn't yet associate with Eastern Europe. Interesting how, in forty odd years, the stern, stout woman in a babushka has been completely supplanted by the stern, lithe supermodel. And how the conditional has also pretty much disappeared. "If he catches that pass," the lazy announcer says seconds after the ball slips through the receiver's fingers, "it's a touchdown." But it was good to embarrass myself in my first teaching performance in front of students I would probably never see again, as much as I might have liked to.

The following day I met my students. The most intimidating were the beginners, who filled a long, narrow classroom with their doe eyes and clear complexions—it's remarkable what an absence of junk food does to teenage skin—and had no idea what I was saying. Luckily, I had learned French in a multinational classroom where the teacher could not resort to English and depended a lot on gestures and theatrics. One fresh-faced student near the front had a long blond braid that the girl behind her affectionately fondled while my words flew over both their heads.

The advanced class was also challenging, as it consisted of only seven students, one of whom, a young man with unexpectedly Asian features, had a severe speech impediment that, combined with his accent, made him difficult for me to understand. During the first class, one of the young women complained that this was the English Language College and that I would be teaching them American English. A few days later I arrived after having mistakenly prepared for the following week's lesson and ended up giving the wrong answer in a tricky exercise, only to be corrected by one of the students. With sweat forming on my forehead, I found the page that contained the right answers. To my credit, I didn't claim my error as a common American usage.

After the final bell, Hania and I went to a cafe with one of her friends from Szymanów.

"How was school tonight?" Zosia asked me in French.

"Terrible," I told her.

"The students didn't learn well?" she asked.

"No, I didn't teach well," I replied.

Somehow, having a conversation in French helped me recover from my bad teaching of English.

Hania insisted that I enroll in conversation classes at the Alliance Française, convinced that it was more important for me to keep up my French than to learn Polish. This carried to a linguistic level a kind of national self-prejudice, or xenophilia, I had noticed in the shops, where the clerks—usually women—would sometimes soften their habitually harsh countenances when I, the obvious foreigner, opened my mouth.

The teacher, Monsieur Nickel, resembled the singer Léo Ferré, with his wild gray hair, and he ran a lively class, one blissfully unencumbered by books, lessons, and exams. Our discussions covered a wide range of topics, even politics. This was French, after all. One morning in early

October, Monsieur Nickel walked in and somberly informed us of the death of Jacques Brel.

I also spent time at the British Council, not taking classes—though my advanced students might have wanted me to—but reading in the library. Teaching three evenings a week, every other day, was for me the ideal schedule. Workdays I spent my mornings preparing, now more diligently than ever, but on my free days—Tuesdays and Thursdays—I could do what I liked. Early on I explored the city, discovering rather quickly there wasn't much to explore. There were no neighborhoods to speak of; there were districts—Żoliborz (where we lived), Muranów, Mokotów, Wola, Ochota—to which Varsovians attributed individual qualities but that appeared to me as mostly indistinguishable collections of drab apartment blocks. The two with true character sat on the other side of the Vistula River: Praga, the shadowy working-class quarter, and Saska Kępa, whose leafy streets and prewar villas attracted members of the foreign diplomatic corps.

The combination of a northern climate and a socialist system meant that there was no street life to speak of, either, not even in the Old Town, whose meticulously reconstructed townhouses, with their reliefs and friezes and fanciful shingles, seemed almost more animated than the people who walked past them. One of the first stops of any visitor to the city was at the history museum on Market Square to see the film of the city's destruction during World War II and its subsequent rebirth, scenes of returning citizens accompanied by strains from Smetana's *The Moldau*. (I always wondered if the communist authorities had decided that an equally stirring piece by Chopin would have been too inflammatory.) Crowds filled the major shopping streets downtown—Marszałkowska and Nowy Świat—but they were purposeful, not spirited, their individual members unengaged in anything beyond their dogged pursuits. And, except for the young women who had gone abroad and enhanced their wardrobes, they were as dowdy as the mannequins in the state department store windows.

The British Council occupied a three-story building on Aleje Jerozolimskie. The second-floor library overlooked the Central Train Station, a modernist structure already showing its age, and the Palace of Culture, a gift from the Soviets that towered over the city in brute, dulled, gloating majesty. When Hania told me the popular assessment of it, that it was "small but in good taste," I felt immediately drawn to Poles, as one always is to people whose sense of humor matches one's own.

The library was the ideal place to read, as I did one particularly gloomy afternoon, Oscar Wilde's *The Soul of Man under Socialism*. "With the abolition of private property, then," Wilde wrote, "we shall have true, beautiful, healthy individualism. Nobody will waste his life in accumulating things, and the symbols for things. One will live. To live is the rarest thing in the world. Most people exist, that is all." I wished I could have taken him on a stroll down Marszałkowska Street.

There were more uplifting works. Reading *A House for Mr. Biswas* in Warsaw, my introduction to V. S. Naipaul, I became the reverse of Graham Greene, who used to take novels by Anthony Trollope on his trips to the tropics. The difference between us was that, instead of retreating back into the familiar in a foreign place, I was escaping one alien world for another, one that in raw, damp Poland appeared immensely attractive. So much so that, rereading the novel many years later, I was shocked by the coldness of the character the author based on himself, a quality that had gotten lost amidst the lush descriptions and opulent vocabulary. Our reactions to books can be determined as much by where we read them as by when.

The other book that captivated me was *A Time of Gifts* by Patrick Leigh Fermor. I must have been attracted by the title—the perfect title for what, I soon discovered, was the perfect travel book—and the cover, which carried a charming illustration of the young vagabond gazing at a large sun that probably reminded me of the sun on the cover of *When the Going Was Good*, which I had purchased in Paris on my way to Provence. I had never heard of the author, but I was immediately intoxicated by his sentences that, in a robustly baroque style, married youthful enthusiasm with rigorous learning—a combination made possible by the fact that he wrote this account of his walk in the 1930s from the Hook of Holland to Constantinople decades after completing the trek. In fact, the book had been published just one year earlier, in 1977. Also appealing was the fact that he was writing about a part of the world most travel writers ignored and I now found myself in: Northern Europe. He made it sound romantic, even exotic, which, in the years preceding World War II, it perhaps was.

Admittedly, the shabbiness of the Soviet Bloc had a certain exoticism, though probably more so if you didn't have to live in it day after day. There was a dreary grandness to Warsaw that gnawed at me: the too-wide boulevards, the wind-swept squares, the lugubrious ministries that made up in bulk what they lacked in height. The city looked as you would expect an Eastern European capital to look—especially one that had been nearly leveled by war—which was like a Washington, DC, badly in need of a paint job. I now understood why Hania hadn't taken to DC the weekend we had visited, why its famous monuments had smacked to her of socialist realism. Because so many of them had been forced on her and her compatriots, she bristled at concrete—even marble—representations of the high-flown.

Though there were clearly two Polands: There was the public one, of puddled sidewalks and drafty post offices and lusterless shops, and there was the private one. Most of the foreign correspondents wrote only about the first, resorting with great frequency to the adjective "gray," since it accurately described the overcast skies, the sodden apartment blocks, the smoky cafés, the drab clothing (on the men at least) and, as soon as fall came, the vitamin-deficient complexions. Either they never entered a Polish apartment—which, considering Polish hospitality, seems unlikely—or they felt their job was not to write about domestic life. Clearly, they had other obligations, though dinner at home with Poles was everything that life on the street wasn't: warm, generous, spirited, amusing. The table was crowded with dishes—often holding foods you didn't see in the shops—and your glass was always filled. You saw Poles as they were, in their element, which was the furthest thing from gray.

I felt privileged to have entry into this Poland through Hania, who could get on a tram and go see most of the people she had ever known in her life—a prospect that to an American adult seemed, even if you discounted the means of transportation, almost miraculous. Since we didn't have a telephone, we'd often just appear at someone's door, where we were always invited in, given a meal, or at least tea and cake. I spoke

French with most of the aunts—feisty anticommunist Catholics—while getting a crash course in Polish manners. A man removed his hat as soon as he entered a building. In the vestibules of some people's apartments, one took off one's shoes and stepped into an oversized pair of immemorial slippers. For greetings and farewells, one started with the women, and always the oldest, working one's way down chronologically. Older men, and some younger traditionalists, kissed women's hands, a gesture that seemed, especially when performed in front of an antique mirror or a prewar portrait, a small act of protest. There are no comrades in a world of kissed hands. At meals, one always kept both hands atop the table. (Hania had always found the unseen American hand at mealtimes suspicious.)

On the street, a man walked on a woman's left side. On buses and trams, if you were lucky enough to find a seat, you immediately relinquished it to any elderly citizen or pregnant woman who appeared in the vicinity. At school, students never addressed teachers with their hands in their pockets. As a rule, all men kept their hands out of their pockets. Poles (Slavs? Europeans?) wanted to see what your hands were up to.

Gatherings with Hania's friends were less fraught with potential faux pas, though on the whole less enjoyable. Most of them spoke English but, not wanting to force my native tongue on everyone, I'd sit in silence while everyone talked. The witless boyfriend. I noticed how Hania more than held her own in conversations; in fact, she often spoke, in her rapid-fire, self-assured way, while everyone else listened, even when the company was predominantly male. For all the courtly deference to women, and the Slavic machismo, there was none of the often-accompanying condescension. After the war, buoyed by the new socialism and the shortage of males, women took jobs traditionally held by men; the two worked side by side and discussed things as equals, especially politics and economics, Hania's forte. She possessed a confidence that was born of intelligence and encouraged by society.

For me it was frustrating, having spent a year acquiring a second language, to be suddenly dropped into another one, especially one that seemed so far out of reach. The language increased the sense of alienation I got from walking the streets and entering the shops. Going to our local *spożywczy*, poorly stocked and rudely staffed, I would recall childhood trips with my mother to our friendly grocer in Phillipsburg. I was a spoiled American, seeing for the first time how much of the world lives. Years later I would read a biography of Ryszard Kapuściński and be struck by the fact that, while he had been off in Africa reporting on the ravages of colonialism, I had been in his country nearly as traumatized by the degradations of socialism. Returning home, we would each find ourselves in Eden. But his, from my privileged perspective, appeared grim and harsh—and winter hadn't yet arrived—its people stuck in a muted half-life.

I was able to find simple things I liked, a useful trait in a travel writer. I enjoyed the trams, especially when I could find a seat and watch the city pass like in an old newsreel. I developed a taste for half-sour pickles; raw sauerkraut (*kapusta kiszona*) that, like the pickles, was plucked from large wooden barrels at the front of our *spożywczy*; Canadian bacon (*polędwica*); and *oszczypek*, sheep's milk cheese from the Tatra mountains that was molded into what looked like a pretty, pointed grenade. I bought beautiful tweed sport coats at Moda Polska, one state store with excellent products, having exchanged my dollars on the black market and become, surprisingly, a man of means.

I also loved the *torby* that hung from the shoulders of Polish males. Most men didn't own cars so they carried their belongings in worn leather satchels that began life stiff and pinkish and, over time, took on a polished, weathered sheen. Mine, stuffed with English textbooks, had already acquired a smooth, mahogany hue, as if soaked in countless cups of tea.

But it was the Varsovians who made Warsaw livable. As a travel writer I have long preached, not just to students but to tourists, the importance of people for knowledge, insight, and emotional connection. Gatherings with family and friends in tiny apartments crammed with books and raucous with discussion revealed to me the true nature of the capital—or at least one segment of its population—and I felt sorry for visitors who were reduced to wandering its joyless streets. Except that even there occasional rewards could be had. People also possess an aesthetic quality that, for the traveler, can have an uplifting effect, something I first discovered in Poland. France is so gorgeous, most of its towns are so charming, that they would stand out even if there were no people in them. In fact, there is a saying—which I don't subscribe to—that the country would be wonderful if only it could be emptied of its inhabitants. Much of Warsaw, by contrast, was unattractive, but on the saddest street, the loneliest bus, you could see the prettiest face. And for a moment, nothing else mattered. In my career I have found myself in places much uglier than Soviet Bloc Warsaw, cities of real poverty and despair (Kapuściński's world), and had it all redeemed, or at least made bearable, by a pair of dancing eyes, a guileless smile, an effortless grace. And of course the beauty is even greater for the squalor that surrounds it.

The interior world was so sacrosanct, the apartment door such a barricade, that I rarely saw our neighbors. So I was surprised, coming home one evening in October, to find people talking excitedly in the stairwell. Inside, Hania told me that there was a rumor that the new pope was Polish. The idea seemed so inconceivable that she could not yet accept it as fact. But that evening, watching the news on television, we learned that the archbishop of Krakow, Karol Wojtyła, had been named the new pontiff. Hania took great pleasure in watching the anchorman as he struggled to hide his patriotic pride and stay not only neutral and objective but also, as an employee of the state-run news service in an officially atheistic country, vaguely dismissive or at least unimpressed. The next day at school I walked into my classes and saw students beaming above copies of *Życie Warszawy* that, I imagined, would not be discarded. The immense pride and delight were obvious, but there was also a sense that life as they had known it was about to change.

On Sunday we watched live coverage from St. Peter's Square. The sight of a religious service on television was unusual enough; that it involved the installation of a Pole as head of the Catholic Church was almost unfathomable. There was an especially moving moment when the archbishop of Warsaw, Cardinal Stefan Wyszyński, a man whose fame had exceeded Wojtyła's, obediently knelt before his old friend and the pope lifted him up from his knees as an equal or even a superior. These two giants of the Polish church, steadfast enemies of communism, embracing at the Vatican in front of the entire world—certainly in full view of all of their compatriots—provided an example of strength and created an image of hope that could not be discounted. Even a clueless foreigner understood that.

Hania, like most Poles, had been raised Catholic, but she was no longer practicing. Yet unlike many Catholic-schooled Americans, even those who are still members of the church, she retained fond memories of the nuns who had taught her. They were, I gathered from her stories, a clever, doughty, eccentric lot, and their strong personalities were accepted, admired, because of their unflinching opposition to the regime.

One weekend a friend of Hania's visited from England. Graham was a professor Hania had met at a planning conference and then befriended, visiting his wife and children in Birmingham. We met him in Łazienki, the most beautiful of Warsaw's numerous parks. Another thing foreign journalists often failed to mention, in their unrelenting studies in gray, was that Warsaw was a surprisingly wooded city, a fact that made me regret I was not more of a nature boy.

Graham, I was happy to see, was old enough to be our father. He had bad teeth and a large nose that supported heavy, thick-lensed glasses. He was funny and easygoing, which I hadn't expected, and he obviously had a special rapport with Hania. She opened the present he had brought her, a small toiletries bag, and then opened the bag, slightly disappointed to find nothing inside it. Sometimes, I was told, Graham hid a banana inside gifts. The sight of Hania searching like a child for a scarce fruit touched me deeply.

A week after he left, Graham sent a letter to Hania. In it, he advised her to come to some decision regarding our relationship, as it wasn't fair to string me along. I sensed that his hope was that she would end it. Hania had told me that once, after drinking, he had tried to kiss her; she had laughed it off as the harmless product of a midlife crisis, and they had remained friends. I suspected that, not being able to have her himself, Graham didn't want anyone else to either, especially an American. He was after all a British academic.

But his summation of our situation hit me hard, for it reminded me that I was at the mercy of a verdict.

Teaching helped keep my mind off things; at school I had not just challenging students but also interesting colleagues. Our conversations went well beyond teaching; in fact, we rarely talked shop and almost never touched on trivial subjects. Sports and pop culture, so dominant in the States, kept a refreshingly low profile in Poland; the lack of junk food extended to the mental diet. TV was boring but it wasn't puerile; bookstores carried classics instead of romance novels; movies, from directors like Andrzej Wajda and Krzysztof Zanussi, delved into history and questions of morality. This didn't produce a humorless people—because of the despised system, there was always a fresh crop of political jokes—but it made them impatient with, or simply uninterested in, the insignificant.

I became friends with Krzysztof Siudyła, who had an impressive command of English and a mordant sense of humor. He had married an Englishwoman, who also taught at the school but on alternate days so there was always someone at home to watch their two daughters. I was in awe of how well, and how wittily, he could express himself in English. In fact, he was funnier and quicker on his feet with my language than I was with it. Though he liked having me to talk to; I could sometimes gift him with an unknown idiom and explain some of the more confounding aspects of American life that he read about in *Newsweek*. My great desire was to watch him teach, as I suspected he was brilliant at it, but sitting in on other teachers' classes wasn't done. Nor was going for beers after class. As soon as the last bell rang, everyone grabbed their coats and headed out to catch their trams and buses home.

After my unsteady start, I was doing OK, but the class of beginners seemed to be missing a lot. All the students were attentive, while the females stood out with their shampooed hair and coordinated outfits. Sometimes, because of the poor selection in the shops, two students appeared wearing the same blouse. The males, by contrast, dressed in a kind of drab uniform of faded blue jeans and green army jacket.

I often prepared for my classes in the library of the US Embassy, which was a short walk from the English Language College. I would use the side entrance on Piękna Street, greet the Marine at his post, and then walk in through the open door. Anyone could do the same; there was no security check or request for identification. Most Poles, however, avoided the place, not wanting to be seen—by whom, one never knew—on the grounds of the American embassy. Those who spoke or studied English tended to frequent the British Council, for it had been situated, sensibly, far from the British embassy and so carried no official governmental affiliation. It was simply, on the surface, an institute of English, just as the Alliance Française was one of French. Of course, it had a small role in the coming global domination of English—though nothing compared to that which the internet would play—but it was free of the stench of propaganda that clung to the library of a Western embassy. I always wondered if our foreign service officers understood this and simply didn't care, or if we needed some help in the art of persuasion.

Naturally, I felt more at home in my embassy's library than I did in that of the British Council. It was like the difference between going to your own church and one of another denomination: The belief was the same but the style of worship was different. With its comfortable blue couches, the American embassy library had a cozier feel; it was a suburban-den library compared to the Council's country-house library. Heightening its homeyness were the nightly broadcasts of the *CBS Evening News with Walter Cronkite*, viewings that were also open to the public. Several-day-old copies of the *New York Times*, the *Washington Post*, and the *International Herald Tribune* hung on wooden sticks by the window, and dated issues of magazines—the *New Yorker, National Geographic, Sports Illustrated, Antæus*—lined the far wall. Perusing the press in the American embassy library was like reading a massive letter from home.

One morning a man walked in wearing a leather vest over a plaid flannel shirt and neat blue jeans. He wore glasses and a mustache that, I assumed, he had grown in the '60s. I silently criticized him for looking so American; France had taught me, for reasons of survival, to try to blend in when living abroad. I now wore my leather *torba* over a grey Polish overcoat; in a few weeks I'd buy a fur hat. But even that would not fool the money changers loitering outside the hotels, who would take one look at my round tortoiseshell glasses and immediately identify me as a potential source of dollars.

A few days later Hania and I were invited to dinner at the home of a young embassy couple, and the man in the mustache was there with his wife. Peter was a poet and a professor from Florida, spending the year on a Fulbright scholarship teaching in the University of Warsaw's Anglistyka department. His wife, Jeanne, was an illustrator who often did filler drawings for the *New Yorker*. They had four children, prompting our hostess to say they were "like Ozzie and Harriet." "More like the Addams Family," Jeanne said, laughing, and I suspected she meant the drawings as much as the TV show. At the end of the evening the four of us headed out together in search of a taxi; Peter grouped us into a huddle to discuss strategies, of which he had none. In France, because I was trying to learn French, I had steered clear of most Americans, but the Meinkes I wanted to see again. And they seemed interested in keeping in touch with us. In addition to shared sensibilities, there were mutual benefits. Peter was a writer; I wanted to be a writer. They were in Poland; Hania was Polish.

On November 1st, Hania and I took a bus to Powązki Cemetery. The crowds grew as we got closer to the entrance, as if we were heading to a sporting event. The air was cold and damp, as it was most days; the her-alded "Polish Golden Autumn" seemed to be a myth. But today the con-ditions struck me as appropriate. There was none of the macabre gaiety of Mexico's Day of the Dead; sober-faced families gathered at graves and purposefully got to work cleaning debris, scrubbing headstones, placing chrysanthemums. Halina Matraś's resting place was so new it needed lit-tle tending. Before leaving, we joined the long lines of people paying their respects at the military graves, candles flickering in glass, smoke mingling with the mist, tears dampening the cheeks of men as well as of women. I thought of Halloween at home, the silly costumes, the cult of candy, and of how Poles appeared to be in touch with the essentials, the eternals, in a way Americans were not.

We were heading helplessly into winter. The sun, which we rarely saw, was now setting in midafternoon. The light had the murkiness of an old aquarium. It made Poles' love for their hamstrung country all the more impressive, though most, even if they remembered the prewar independ-ence, had never experienced a sunny Advent.

I was used to winter, I thought, but got to taste it full blast while waiting for trams—trams that ran irregularly, especially, it seemed, when the temperatures dropped. It occurred to me that, for all the time I'd passed in the cold, very little of it—I think of the rare football game—had been spent stationary. I had walked, or skated, or shoveled, and then, usually, entered a house or climbed into a car. Now long minutes passed with me standing, dependent on public transport, defenseless against the frigid gusts. Yes, I now had my new hat—made of nutria fur—but, trying to appear as tough as Poles, I refused to untie the knot at the top and release the earflaps. This way it kept its proper politburo shape.

The week before Christmas, I taught my classes English and American carols. ("The Twelve Days of Christmas" worked as an excellent exercise in pronunciation.) While there were few civic decorations, private celebrations of the season were not discouraged, probably wisely in a country that had been officially socialist for three decades and overwhelmingly Christian for more than a millennium—and that was now the homeland of the pope.

For Christmas Eve we were invited, perhaps out of pity, to the home of a distant relative who held a high position in a large state enterprise and was, everyone assumed, a member of the communist party. The apartment was in the south of the city, in the same vast, dehumanizing complex where Krzysztof lived, so our host's party affiliation, if it existed, had gotten him no housing perks. I would have preferred my new friend's Anglo-Polish Christmas, but Hania appeared nonjudgmental about this branch of her family, which seemed equally at ease celebrating the holy birth. Poles had learned to make accommodations, and those who didn't had learned to accept those who did, especially when they were connected by blood. Also present was a cousin of Hania's who worked as an editor at *Sztandar Młodych*, the paper where Kapuściński had gotten his start.

We arrived around three. The Polish *wigilia* traditionally begins when the first star is spotted in the evening sky, which in December can be in late afternoon. Before sitting down to eat, we each took the *opłatek*—a rectangular wafer similar to that used for communion—that had been placed on our plates and went around the table offering a piece to everyone in attendance, along with wishes for the coming year. This lovely ritual lasted a while as the exchanges, far from being perfunctory, were detailed and individually tailored before ending with the obligatory three kisses on the cheeks.

When we were finally seated, two types of soup were served: cream of mushroom and clear borsht with small dumplings filled with finely ground mushrooms. Then came fish in aspic, pickled herring, and carp, a holiday staple that, not unlike turkey in the States, is rarely eaten at any other time of the year. No meat is served at the *wigilia*, earning the meal—which, according to custom, consists of twelve courses—the title of a fast. I filled up on boiled potatoes, the only type of potato these potato-loving people seemed to make.

After eating and opening presents, we headed out to the cathedral for midnight mass. Stately pine trees flanked the altar, and worshippers filled every available space. I heard, for the first time in a sanctuary, the exquisite Polish carols, including the achingly beautiful *Lulajże Jezuniu*, echoes of which appear in Chopin's Scherzo No. 1 in B Minor, op. 20. At the end of the service there was none of the facile bonhomie I was used to from home; people headed off into the night with serious, contemplative expressions. They were still in thrall to the mystery and, because of the system, unused to lavishing good cheer on strangers.

The next day, with no invitations and little food in the apartment, Hania and I took a bus to the Hotel Europejski. The dining room had an institutional feel; the lighting was harsh; I was homesick; and Hania, with no relatives around, felt the full force of her first motherless Christmas. We rode a silent bus back to Żoliborz.

A few days later a storm arrived that would stamp the season as "the winter of the century." Temperatures dropped to well below freezing and brought with them a torrent of blizzards. I had always liked snow, the way it transformed the world and brought life to a temporary standstill; I had inevitably been saddened whenever it stopped, forever disappointed that the accumulations weren't greater. I had spent twenty-odd winters hoping for the snow of the century.

Just not in Poland. Everyday existence in Warsaw was difficult enough that I didn't need nature to refine the experience. Apartments were toasty—at least they had been until this storm started pounding the windows—but what if the heating failed? Our local *spożywczy*, never flush in the best of times, might not see new shipments for weeks. One afternoon I sat by the window and watched with dread as wind-whipped snow obliterated everything, including the belief that I was in a European capital. All I could see was a howling white emptiness.

It produced in me a feeling of helplessness I didn't communicate to Hania. I didn't want to appear weak. But I had never felt so strongly the power of nature, and I was experiencing it in a place that seemed dangerously fragile, on the edge of a great wildness. Of course, people had lived here for centuries, in much more primitive conditions; they had fought battles, run off invading armies in winter. But they had not grown up in 1950s America.

On a snowy, blustery New Year's Eve we headed out to a party at the apartment of Kasia, one of Hania's university friends, and her husband Krzysztof. I had never seen so many smiles on buses. True to their forebears, Poles, at least the young ones, were reveling in the storm. Their monotone life had been turned into an adventure. Years later, Krzysztof from school would tell me of the Polish psychologist who wrote that Poles can be happy only in those situations when they have no reason to be.

On January 1, 1979, a national disaster was declared. It seemed an ominous way to begin the year. We also learned that the airport had closed, news that added to my feeling of disquiet.

I of course had not envisioned any of this that day in London when I had met Hania. And for all my discomfort—physical and psychological—I didn't regret coming; I was grateful, sort of, for the new experiences. (Someday you'll look back at this and write.) And I was more in love with Hania than ever. One rare, sunny afternoon, meeting her in town, I watched as, in an equally unusual burst of athleticism, she ran down a snow-packed alley into my arms.

Toward the end of January, we nervously made the trip back to Krucza Street to apply for a visa extension. The man who had gifted me the original visa was still there, a chunky functionary in a thick wool sweater and with curly blond hair. He didn't look very happy to see me.

Returning a few days later, we learned that my request had been denied. Hania wasn't too concerned; when refused favors from the authorities, you waited a bit, she told me, and tried again. You might find your man in a better mood or chance upon someone new.

On my second try, we again encountered our sweatered blond man. This time, too, I received a denial.

We went to dinner with Peter and Jeanne at the Cristal Budapest, the best of the city's few ethnic restaurants. Each one represented a communist country, and, while this could have meant great Chinese and Cuban food, it didn't. Those countries were too distant and exotic for faithful re-creations of their celebrated dishes, and few Varsovians dined at Hawana or Szanghaj. Hungary, by contrast, was not only in the same bloc but was also one of the few countries there that Poland historically had close ties with. It helped, I always thought, that they no longer shared a border; the Germans and Russians were both despised—for obvious reasons—while the Czechs were mildly ridiculed. Many Poles traveled to Budapest to celebrate New Year's, and a perennial at parties, along with vodka, was Egri Bikaver, the Hungarian red wine known as "bull's blood." For my part, I liked Hungarian cooking because it took a lot of the bland things northern Europeans ate and spiced them up. But I wondered why, given the bias toward communist cuisines, a city like Bologna wasn't represented. I could have gone for some tagliatelle in meat sauce.

Over flame-heated bowls of goulash, we told our friends that I might soon have to leave the country. Jeanne reacted with shock and concern, while Peter appeared to take the news in stride, like a man familiar with the unsparing ways of the world.

On our third trip to Krucza Street, with three days left on my visa, Hania was kept downstairs while our man escorted me upstairs. This was new; she had always stayed with me to serve as interpreter. I was led into an office where a large man sat behind an empty desk. He summarized my situation in heavily accented English and then said to me, "If you help us, we can help you."

I hadn't been expecting this, a scene from a movie. I asked him to explain what he meant.

He said that I was in a unique position, having contacts with Poles and contacts with foreigners. All I had to do, he claimed, was come to his office once a week and tell him what these people were saying. He noted that it was not that different from the journalism I had been doing in the States. I found the comparison insulting but, reading years later of Kapuściński's cooperation with the authorities, which helped earn him those long stints abroad, I finally understood my apparatchik's thinking.

I told him I was interested only in teaching, work I had been doing for less than a year. I felt a bit hypocritical claiming the profession as my new passion when my passion sat downstairs. The man knew all about Hania and clearly hoped that I would do anything, including work as an informer—a word that was never uttered—in order to be with her. If I accepted his offer, he said, my visa would be extended indefinitely. But I wouldn't be able to tell anyone, including Hania, of my new avocation. This was the flaw in the proposition, particularly for people who, unlike me, felt no political or patriotic allegiance. In order to remain with your beloved, you had to betray your beloved.

After several minutes, I told the man what I had felt from the start: that I wasn't interested in his offer. He showed no emotion—I'm sure he had heard my answer before—and said that there was nothing then that he could do about my visa.

I had received my verdict, just not from Hania.

I walked down the stairs and found her waiting. I didn't say a word, fearing repercussions if anyone heard me reveal what had just happened. I led her outside and walked until the noxious building fell out of sight. Then, finally, I revealed the reason I looked so shaken.

She was not impressed. Friends of hers had had similar interviews when requesting their passports to go abroad. She jokingly asked what the salary would be.

At school, I told Mr. Kuczma. "It all adds up to your experience," he said, before assuring me that I would always have a job if I ever returned. It seemed unlikely, but I appreciated the kindness. Now there was a school in Poland and a farm in Alsace where I could always go. My fellow teachers reacted with disgust but not surprise. My students learned only that I had to leave Poland. (Mr. Kuczma had warned me at the start that you never know who is in your classes.) The advanced class, including the woman who'd expressed concerns about American English, seemed genuinely disappointed. One of my beginners, walking out after class, whispered as she passed me, "'Tis a pity."

Once again, I was leaving a city in a way I hadn't anticipated. The plan now was that I would go to Greece—I had a friend in Athens who had been at school with me in Aix—and try to find a job till the spring, when I would attempt to return to Poland for vacation with Hania. By then, with luck, the snow would be gone.

I reserved a sleeper on the overnight train to Budapest. I packed my heavy, soiled leather suitcase. The *złoty* had no value outside Poland so with my money I bought a gold ring. Hania gave me her father's old watch, so I'd have an extra in case mine stopped.

On a cold February evening, Hania and I got a taxi to the Central Station. Descending to the platform, we found my train and quickly learned that no sleepers were available. We showed my reservation to the conductor, who remained unmoved, even after seeing I was obviously a foreigner. Hania and I kissed goodbye, and I resignedly boarded. A railway worker appeared; Hania immediately collared him, oblivious now to my wave of farewell. She was still badgering him as the train pulled away.

CHAPTER 3

Greek Tragedy

A FUZZY WHITE BALL OCCUPIED MY COMPARTMENT. THE STANDARD winter headwear of the middle-aged matron, it covered, in hirsute mimicry, the hair of the passenger lying on the middle couchette. A well-trained male, I removed my fur hat, despite the chill, and placed it on the luggage rack.

As if in an instant, the train emerged from the city into the countryside. I took the couchette opposite the hatted woman and stared out the window at mile after mile of untrammeled snow. A storm a few days earlier had smoothed over the previous accumulations, and I wondered how many flakes had had to fall to cover so deeply so much earth. We seemed to be the only movement and light in that frozen land. I had a vision of an entire nation asleep under snow, mute and immobile, waiting the long wait until spring.

Then I closed my eyes and tried to sleep, but my mind wouldn't let me. Now I was like Theroux, off on a railway adventure, but with a heavy heart. I had been not only separated from Hania but prematurely pulled out of Poland, after a far from successful stay. I had struggled with everything: the language, the weather (having experienced the two darkest seasons), the life drained of amplitude. And Hania was aware, if not tired, of my frustration. I had become critical of the way things were done. Thinking of all that I was missing in the West made me moody and frequently whiny. None of my friends had experienced what I found myself in, and, instead of appreciating my situation, I often longed to be back with them. Unlike my summer in Alsace, six months in Poland had made me a flawed traveler.

I dozed off finally and was awakened around midnight by a sudden burst of light and a barked demand for passports. We had reached the border. In the harsh glow of the rudely illuminated compartment I saw the fuzzy white ball rise from its pillow to greet grim officers. Happily, no one found my two watches suspicious, or desirable. After an hour of inspection, or perhaps simply malice, the train slowly headed off into Czechoslovakia.

In the morning, the snow was gone. We pulled into a beautiful old station, its iron-and-glass shed soaring like a royal welcome back into grandeur. I felt a surge of groggy excitement; I was lovesick, displaced, deprived of sleep, and filled with regret, but outside a new city awaited.

Already walking down the platform I saw something I hadn't seen in months: purchasable sandwiches. Small rolls filled with cheese, tomato, and hard-boiled egg appeared behind glass like a sudden specter of prosperity, fruits of the country's so-called Goulash Communism that incorporated elements of free-market economics. I stored my suitcase, changed some money, and then checked evening departures: the trinity of tasks that, in the coming days, I would perform with alacrity.

There are few experiences more pleasurable for me than walking out the doors of a train station into a foreign city. I feel an excitement similar to that of the mountaineer on his peak, exhausted by the effort of reaching this ground but exhilarated by the expanse that lies at my feet. With five hours remaining until my night train to Bucharest, I picked a street that looked as if it might lead to the river and walked, not planning to see anything but, with the optimism of the *flâneur*, hoping to see everything.

Budapest was my first Eastern European capital after Warsaw, and it looked at once brighter, older, richer, and more Western. A new supermarket featured a row of orange cash registers beneath a ceiling of colored streamers; next door, a sidewalk beverage dispenser offered a delicious concoction of malted milk shot through with orange strands. (It had been months since I'd tasted citrus.) Just walking sure-footedly on ice-free sidewalks was a great novelty.

Quick-service restaurants sold broiled chickens, meat pies, French fries—a beautiful sight after months of boiled spuds—and bulbous sausages sizzling in the roseate glow of heat lamps. Closer to the river, modern travel offices, boutiques, and cafés sported plate glass windows and neon signs. Stepping inside a restaurant, I watched a thick-lipped, raven-haired waiter in a kind of eternal shrug shuffle with gravyed plates across a sawdust floor. Like the big-windowed boutiques, he could have been plucked from midtown Manhattan.

I stayed in Pest the whole time, looking across the Danube at the hills of Buda. Trams ran along the river, and a tall Intercontinental Hotel rose beside it, a building so ugly and imposing that, in representing the West, it appeared as an aesthetic, and an intentional, argument against it—as if the local authorities had allowed the chain's presence in the city only if it built the most hideous structure.

In one narrow street I passed through a gauntlet of women selling versions of the red-and-black patterned vests and skirts that they had on. A vision of the village in the middle of the city. Poland also had a rich folk culture—the song-and-dance troupe Mazowsze traveled the world—but, except for the state-run Cepelia shops, one never saw manifestations of it in the capital.

Around dusk, I made my way back to the station. I had no map or guidebook; I simply relied on my well-developed urban sense of direction.

The train to Bucharest appeared like a preview. Garbage spilled out of the small metal container beneath the window and filled the compartment with a piercing stench. One of the seats had fallen, putting its occupant—a long-haired Bulgarian in a new denim suit—literally on the floor. Slightly higher up sat a Hungarian who looked at the start of this overnight journey the way one usually does at the end. Next to him sat a Romanian peasant whose wife sat across from him and next to me. We were soon joined by another young Bulgarian man.

The Romanians and the Bulgarians each talked amongst themselves. I listened closely to the Romanians in futile attempts to hear some similarities to French. As if sensing my disappointment, the peasant woman showed me photographs of her children.

Our stop at the Hungarian-Romanian border took over an hour. It was my third border crossing in two days, and already the process was becoming an inseparable part of my journey. The customs and passport control officers, in their military-style uniforms, were barely distinguishable from country to country: a slightly different shade of wool fabric, a partial change in complexion, a newly situated red star.

My identity, if not already known to my compartment mates, became so at borders. The Romanian passport officer looked at everyone else's passport and returned it; mine he kept and took off somewhere. After twenty minutes he had still not returned. I wondered what he could possibly be checking—and how. The peasant woman, seeing my consternation, pointed at each person in the compartment, followed by the word "socialist." Then she pointed at me and said "capitalist."

She asked the Bulgarian how much he had paid for his new denim suit. Unable to make himself understood in his native language, he took a pen and started writing the amount on the palm of his hand. The Hungarian quickly proffered a piece of paper. Then the woman turned to me and asked if I had any jeans. There could have been no clearer indication of the acquisitive nature of the human species.

My passport was eventually returned, and the train was allowed to continue its journey. At Cluj, the Bulgarians departed, and, shortly afterward, the Romanians did the same. The Hungarian and I each stretched out on the newly emptied seats.

When we awoke, it was to the snow-covered pine forests of the Transylvanian Alps. Villages appeared, fretworked clusters of tall wooden houses in pastel hues. Surrounded by snow, they looked as soothing as butter mints.

In Bucharest, I tried to reserve a couchette on the night train to Sofia, but none was available. At least not for me. The city, just two years after the earthquake that gave Nicolae Ceaușescu a pretext for his "modernization" plan, looked officious, sinister, and bleak. Amidst the gray apartment blocks some elegant villas survived, one of which housed the US Embassy. The street in front of it was blocked to traffic.

"What's going on?" I asked a man I'd seen coming out of the entrance.

"Oh, it's always like this," he said before hurrying off down the street. I was missing Hania terribly and, under the circumstances, the bed we had had to make every night. But this was the first time I longed for Warsaw.

Late that afternoon I entered a restaurant and joined three students for pizza and wine. Throughout the meal, they raised their glasses and wished me "*bonne santé.*" A beggar woman approached our table and was quickly shooed away by the waitress. I looked at my watch and saw that it had stopped; luckily, I had a spare.

Outside, darkness had already descended. A series of shadowy, indistinguishable forms passed beside a concourse of muddy cobblestones. I gave a vendor a few *lei* and received a disc of fried dough that, when I bit into it, oozed melted cheese.

Inside the dimly lit station, I walked past hollow-cheeked men in high astrakhan hats. No one carried a suitcase; instead, they lugged bales or bundles or packs draped across their backs or slung over their shoulders. I was the only person wearing glasses.

The waiting room contained rows of swaddled bodies slumped on straight-backed wooden benches. The air was thick, though no one seemed to be smoking. Children in tattered clothes ran around a metal crutch on the floor. The atmosphere of tedium, endemic to most terminals, was here compounded by want and oppression. An image came to me of my parents, sitting safe and warm in our well-lit den in Phillipsburg (even though there it was still afternoon), and with it a sense of how far I had traveled. I was in a place they could not imagine.

Which didn't help me shed my immense uneasiness. Nighttime stations are rarely pleasant, and this one had the murk of a netherworld. I had never felt so vulnerable, so removed from the known. I was on my own completely, as all solitary travelers are, of course, but I was such an obvious outsider, someone with no reason to be there. I could have been robbed, or beaten, and no one would have had any reason, other than basic human decency, to come to my aid. I waited for my train hoping no one would harm me because in a way—with my glasses and watches and American passport—I felt I deserved it.

It helped that I was in an authoritarian country, where punishment is harsh for all criminals except those working in the government.

The night train to Sofia was the least populated of any I'd been on. I shared a compartment with a young man heading home from his job in Prague. In two days I had met more Bulgarians than I had in the previous twenty-six years. He didn't care for Czechoslovakia, he said, but the money was good. On hearing that I was American he left the compartment and returned shortly with a friend who, politely, asked to buy dollars.

The train stopped at the border, which here was a river: the Danube. It had followed me all the way from Budapest.

First the Romanian officials boarded.

"Pennsylvania?" the passport officer said, reading aloud the place of my birth. "Heh, is Romania."

Far from it, I thought.

He was replaced by Bulgarians. One took my passport; another asked me to open my suitcase. On a transit visa, I was not required to, but I didn't want to be difficult.

"Any pornography?" the young man asked in a tone of veiled hopefulness.

My compartment mate and I were standing at the window in the corridor when the officer returned with my passport. He handed it back to me, said something to my companion, and departed.

"Did you understand?" my companion asked. "He says me socialist, you capitalist. Is not good for me to talk to you. Not good company." He seemed little bothered by it at the time, but later that night he departed, and I slept alone. Albeit sitting up again.

When I awoke, I found four new people in my compartment. I checked my good watch for the time and saw that it too had stopped.

We arrived in Sofia, where the train to Greece was to leave shortly. It was a ten-car train, the first car of which was the only one going all the way to Athens. In it I found four cheerful young Polish women from Poznań, in wool sweaters and tartan skirts, surrounded by a phalanx of fawning Greek men. Tapes of bouzouki music played while glasses of vodka and ouzo were hoisted. This festive meeting of the Baltic and the Mediterranean seemed to signal the end of my dour rail odyssey.

Outside, southern Bulgaria passed under low gray skies. I had been persistently moving south, but not far enough for the sun to appear. A muddy river flowed along brown, growthless hills, calibrated like steps from top to bottom. Rustic houses of unsmoothed stone rose at the edges of fallow fields. Along the tracks, shepherds and peasants warmed themselves around fires. During station stops, obliging locals took our empty bottles and filled them at old stone fountains. I'd read about the deliciousness of Bulgarian water in *Foreign Faces*, that fateful book.

Two mutts greeted us at the Greek border. There must have been a customs inspection, but if there was it was so perfunctory it didn't make an impression. But the first stop did. The yellow station house wore a fresh coat of paint, and the tree in the courtyard was neatly trimmed. Beside it stood a wizened, extravagantly mustachioed man who soon entered our car bearing bags of peanuts and pistachios, oranges, halva, sesame candies, and Turkish delight. When we were moving again, a younger man marched through the car ringing a bell and offering souvlaki, crusty bread, and cold beer. At the next stop, a squat woman came on with dustpan and brush and quietly swept all of the compartments. In thirty minutes I saw more efficiency and service than I had seen in the last four days—or perhaps six months. I had arrived in a new country, but I had returned to a familiar prototype. And, finally, the sun was shining.

Contrasting with the suddenly satiny journey was the growing ruggedness of the landscape. Mountains rose in the distance, and sheep and goats grazed in rocky fields. Groves of olive trees sometimes softened the scene, as did the occasional white-domed chapel flanked by proverbial dark-green cypresses.

The tartaned Poles departed at Thessaloniki, where I was joined by a middle-aged couple. The man asked in broken English if I was married. I shook my head and tried to explain the presence of the ring on my finger. He didn't understand, and we eventually fell into silence.

About an hour later, he asked me for the time. I had removed my gray Polish overcoat—for the first time in days—and my watches were visible on my wrists. I said I didn't know and lifted my arms to show that both of them had stopped: the Bulova Accutron that had been a high-school graduation present from my grandmother and the prewar Swiss timepiece that had belonged to Hania's father. After learning that I was wearing two worthless watches and a bogus ring, the man left me alone for the rest of the journey.

We arrived in Athens the following morning. With a mixture of relief and uncertainty, I disembarked from the Iron Curtain Local and headed out into the Cradle of Western Civilization.

At the station I stored my suitcase and changed some money, then found my way to Syntagma Square, the only person in the city wearing an overcoat. I bought a gyro from a sidewalk vendor—tasting in its homeland the sandwich I had grown so attached to in New York—and its dripping grease stained the hem of my coat. It felt like a christening.

The address I had for Christos Kontovounissios, I was pleased to discover, actually existed, for when I showed it to people I received directions instead of shrugs. My love of walking around new cities was all the greater after five days on a train. The mild weather and look of prosperity—well-stocked shops, bright advertisements, outdoor cafés—cancelled out any disappointment I might have felt at the unprepossessing architecture. I was back in a prosperous, al fresco capital.

Christos lived with his sister in a modern, low-rise apartment building relieved, like the neighboring ones, by long balconies with low-slung awnings. Small of stature, he had the same leonine look I remembered from Aix, though his brown mane had been considerably reduced and his facial hair trimmed into a Continental goatee. His blue eyes still telegraphed compassion and an intense, earnest desire to comprehend.

We sat on the balcony, near a lemon tree in bloom. "Winter is over," Christos announced, uttering three words I had never heard in February.

I told him of my experience in Poland, the reason for my sudden departure, and my desire to find a job until spring (real spring) when I hoped to return to Warsaw. He told me of an English-language institute, with schools throughout the country, where I could probably land a job.

Christos, after leaving France, had studied diplomacy, and now was working for the Greek foreign ministry. He was still attached to Linda, the American girlfriend who had been with him in Aix and who was soon going to be moving to Greece. We had been foreign students together and now we were both wooing foreign women, the usual difficulties of romance complicated by geography, bureaucracy, and decisions of immigration.

Before paying a visit to the language school office, I tried some other job possibilities, including at the American embassy. Ideally, I wanted to stay in Athens. In the evening, Christos and I would sometimes meet downtown, commandeering a table on the sidewalk as office workers walked arm in arm, sometimes hand in hand, and boy waiters swung their coffee deliveries on silver trays. No one seemed to be in a hurry. Adding to the almost Levantine languor was the music that blared from shops, all of it Greek, as if no other music existed. In communist, "Western-hating" Poland you could not go a week without hearing ABBA. I began to understand what people like Durrell and Leigh Fermor saw in the place.

I did the obligatory sightseeing, visiting the Acropolis and roaming the souvenir-draped streets of Plaka. Some days I stayed home and read *Of Human Bondage*, which I had found among Christos's paperbacks. I was astonished by how much Maugham knew about my old relationship with Molly.

One day, after another unsuccessful search for work, I returned to the apartment and found Christos listening intently to the radio. A revolution, he said, was taking place in Iran.

I finally went to the office of the language institute. A man sat at a desk in front of a large map of Greece. Without even looking up, he asked, "English? American?"

"American," I said.

He stood and, pointing to a dot on the map far in the north, said, "We have a job for you in Arta."

It was my first experience, though I didn't yet realize it, with Greek anti-Americanism.

The next morning I went to the train station and retrieved my huge leather suitcase from a large room filled with luggage. Though it had no lock and had sat for over a week, it had not been touched. Then I made my way to the bus station.

After a seven-hour ride I arrived in a town about as far from the islands and travel-poster Greece as one could get. A film of dust seemed to hang in the air. I lugged my suitcase past black-robed widows and an Orthodox church with a communal oven behind it. The only similarity to Athens was the bouzouki music emanating from shops.

The school stood on a street with a church and a movie theater. If he was happy to see me, the director did his best not to show it. After a few minutes I discovered the reason for his cold reception: Despite the Cambridge Certificate of Proficiency in English hanging on his office wall, he spoke very little of the language. He presented me with my schedule, which included one class on Saturday evening, effectively quashing any chance of my going away for the weekend and giving me, I suspected, my second taste of anti-Americanism. Then he handed me some old English textbooks.

We walked to his house, where his wife served moussaka and I met the plumber in whose home I would be staying. This man spoke no English but had a cordiality that was totally lacking in the director.

The plumber lived with his wife and two small daughters in an old stone house halfway up a hill. I was taken to a room in the front where a single bed had been pushed against a wall. The warmth outside had not penetrated the edifice, and the room was chilly in midafternoon. I wondered what it would be like at night. The house of course had no heating, this being sunny Greece. The bathroom, reached through the kitchen, had no tub or visible shower. Not that I was eager to undress in these temperatures. And the door of the bathroom hung open a few inches, allowing every sound inside to be heard in the kitchen.

Getting ready for bed that night I could see my breath. I slid between the icy sheets and burrowed underneath the blanket, atop which I had placed my Polish overcoat. Gradually, the incipient interior warmth spread to the fingers of my exposed hand.

Awakening the next day, I initiated a routine that I would reenact every morning for the next two months: I jumped out of bed, pulled on my sweater and cold jeans—the sensation not unlike that of putting on a damp bathing suit—and, after passing through the kitchen and quietly using the bathroom—a sheepish *"kalimera"* to the family seated at the breakfast table—I dashed outside to sit on a sun-warmed bench.

My students, I discovered later that morning, were much younger than the ones I had had in Warsaw; most of them were still in grade school. There was a collective gasp as I wrote my name on the blackboard; they had never been taught by anyone left-handed. I tried to explain that the condition was normal, at least in some parts of the world, but no one had any idea what I was saying. I was an unintelligible freak, their new teacher.

They, for their part, were incredibly cute, despite their antsy behavior and their incongruously august names. Socrates was an enthusiastic nose picker, and Aphrodite showed little passion for anything other than

sunflower seeds. They seemed not only restless but insolent until I learned that the dismissive "tsk" combined with a scoffing lifting of the head was the Greek way of saying "no."

The textbooks were British and shamefully outdated. The word "wireless" was used instead of "radio." It was strange to think that, back in Trenton, my former colleagues were now writing on word processors. At the end of one chapter, a review question asked, "In Africa, whom do black men obey?" As shocking as this was, it made me think of the joke I'd heard in Poland, which was never attributed to its supposed creator, John Kenneth Galbraith: What is the difference between capitalism and communism? The first is the exploitation of man by man while the second is the opposite.

The textbooks sat on shelves in the director's office with a few orphaned paperbacks: Evelyn Waugh's *Men at Arms* and an English translation of Nikolai Gogol's *Dead Souls*. I couldn't use them in class, but I could take them home, since nobody else had any interest in them.

And so I began a new life, one that seemed sadistically designed to show me how good I had had it in Poland, a place that—thanks to the postal system of a socialist country working in tandem with that of an antique one—seemed almost insurmountably distant. The wait for the first letter from Hania was excruciatingly, self-doubtingly long. Once again, I didn't speak the language, and those who spoke mine in Arta were extremely few. Fellow teachers, Greek men all of them, had families or other jobs; they taught their classes and then left the premises. The director had instituted no communal tea or coffee, and the rare presence of an American—at least an American man—held no attraction.

There was an American woman in town; she had married a local, and we would sometimes meet on the main street during the ritual evening promenade. But we could talk for only a few minutes as anything longer, she explained, would be a sign to the population that we were having an affair. So I would stroll by myself, marveling at the ease with which people let papers, wrappers, napkins fall to the ground and then continue on their way. Later in my stay, when I had lost all hope of acceptance and sense of embarrassment, I picked up a paper a man had jettisoned mid-stride and, catching up to him, presented it triumphantly with the words:

"You dropped this." The foreign tongue only partially explained his look of bewilderment.

After the promenade, I would repair to a café—as one of my textbooks would have put it—and order an ouzo. Liquid licorice. It came with a tall glass of water—which I used for mixing, enjoying how the two clear liquids combined to form a cloudy one, like two pristinely argued defenses suddenly making the listener doubtful—and a plate of food. The victuals varied from café to café and from day to day. Sometimes there would be olives; other times, peanuts. Occasionally you'd get lucky—cubes of cheese and bread—or even luckier: octopus tentacles. With each new drink came a new plate. It was an impressive custom, born not out of a marketing strategy but a familial ethic of hospitality and plain common sense. I never saw a Greek man walking home drunk.

I say "man" because the cafés were frequented only by men. This was a surprise to me, not just as an American, but as one who had arrived from Poland. In the evening, Greek wives cooked and cleaned and got the children ready for bed while their husbands pontificated in the cafés. As it had been for centuries, so it was now. The place was as antiquated as my textbooks.

Seeing this unquestioned purdah in the latter half of the twentieth century, in a Western European country, would have depressed me even more had I wanted to meet women. My obsession with Hania rendered that idea moot, not so much from a moral standpoint as from a psychological one: She so dominated my thoughts that I had no room for anyone else.

Few people have ever nursed a drink as I did in Arta. Every night, because of the uninhabitable condition of my room, I used the cafés as my office—years before the first laptop invaded Starbucks. Writers had long made the coffeehouse their home away from home—in Paris, Vienna, Budapest—but more for socializing than for working. The café was a (male) meeting place in Greece as well; I was the only person in Arta who spent his evening outings alone, and the only one who sat at a table with pen and paper. And if there *was* another, he didn't hold his pen in his left hand.

It was a pleasure to write in longhand again. I composed long letters, mostly to Hania, but also to my parents and friends, including Peter and Jeanne. The tone was usually humorous, since humor seemed to be my strength and I was now living amongst people to whom it was invisible. And it allowed me to describe my plight—the surprising cold, the rambunctious students, the incompetent director, the overall backwardness—in a way that wouldn't sound like whining. Tragedy doesn't always need time in order to become comedy; comedy can be born swiftly out of a desire to present an attractive self-portrait.

If my misery made loved ones laugh it was not in vain. Though my humor was often sophomoric—I made callow jokes about electrolysis—and strained. Because the cinema near the church and the school frequently showed adult films, I wrote that the street offered "titillation, salvation, and pronunciation." Hania learned that I no longer needed to move my toothbrush in the morning because my teeth were chattering so violently. The irony that my suffering from the cold was far greater in Greece than it had been in Poland—never spend winter in a famously warm European country—did not justify my complaining to her about it. And I think all of my correspondents learned of my plan to run for mayor on the slogan, "Arta for Arta's sake."

I was witty (or not) in one-way, written conversations—many writers' favorite form of communication. The missives, while giving me something to do in the evening, and allowing me to show off—or, more accurately, work on—my writing skills, also served as helpful reminders to myself, in the loneliness of Arta, that I had friends, family, a girlfriend out there in the world. Hours would pass, patrons would come and go, waiters would wipe the tables, and my pages would fill with words describing my current life—sometimes the very café in which I sat—while transporting me to my past lives in, depending on the recipient, Phillipsburg, Washington, Kutzenhausen, Trenton, Warsaw. The morsels on my plate would have long ago disappeared; the level of ouzo in my glass would drop almost to empty. I loved the taste, the strong sweetness of a flavor I had always associated with the color black, and the almost sacramental quality the well-spaced sips gave to my public labors. I never could afford more than one or two. One night, in a café I rarely visited, so much time passed between my ordering the drink and my getting up from the table that I departed without paying the bill. Outside I realized my misdemeanor—the cold air instantly clearing my head—and, when I went back inside, the owner waved me away with a fatherly look of forgiveness.

When I wasn't writing letters, I worked on a story about my train ride. It was the most traumatic trip I had ever taken, and writing about railway journeys was clearly in vogue. Unfortunately, I was still more under the influence of Benchley than of Theroux. The tone I was using in my letters—one of blithe exaggeration, to produce laughter instead of pity—was wrong for my story, which needed to convey the reality of the experience—all the discomfort and tension—especially now that I had survived it. Writing flippantly of a miserable situation while you are in it has commendable elements of self-deprecation, but doing so after the fact risks rendering the account inaccurate or, at the worst, dishonest.

In the morning, I continued my writing on my bench. Within minutes the sun would steal the chill from my jeans, warmth would gradually return to my hands; eventually my thick green sweater would come off. It was like sitting by a hearth, early in the day, the sun a necessary—and happily constant—source of heat. I never thought about skin cancer. The idea of putting on a hat, even if I had had one, would never have occurred to me. I didn't, nor did anyone else in the town, wear sunglasses. Though I was the only one who basked in the sun, just as I was the only frequenter of nighttime cafés, a traveler between worlds of light and dark. There was nothing contradictory about my behavior; it was simply my survival tactic for living in an unheated house during a northern Greek winter—in a town where I didn't have a friend.

When not writing or preparing the day's lessons—which never took very long, as my teaching had evolved into a routine of mostly improvisational theater—I read Waugh and Gogol on my bench. Waugh's taut, assured, expertly cut sentences, even when used to depict a man's world (because of my bleak circumstances I was more in the mood for the cavortings of the Bright Young Things) reminded me why I wanted to be a writer. Gogol, for his part, was a delightful revelation, transporting me to nineteenth-century Russia with a humor and sensibility that seemed very modern, certainly more modern than the school's English textbooks.

There in the sun, I also drew cartoons and caricatures, my new surroundings, in all their bumptious Mediterranean richness, providing me with countless subjects. I had never been so creative, and the fact seemed to suggest a dispiriting truth: that sustained artistic productivity results from unhealthy, self-imposed solitude.

Other than a few basic words—*kalimera, kalispera, efharisto, poli kala*—I made no attempt to learn the language, which went against my principles as a conscientious traveler. But Greece was a pit stop, a temporary shelter, a place where I had hoped to sit out the rigors of winter (ha!) before continuing on with my life. It hadn't been a part of my original plan; it had appeared suddenly as a last resort. I had arrived with a certain amount of enthusiasm but quickly realized, after landing in Arta, that this was not a place for me. Which was just as well, as there was no shortage of writers, including some of my favorites, who felt very differently and had, by their attentions, roundly excluded the country from the ranks of the unsung. So, it seemed pointless to try to learn Greek. Instead, some mornings on my bench, I studied the language of a less glamorized country: *dzień dobry, dobry wieczór, dziękuję, bardzo dobrze.*

My first escape from Arta came on a weekend when, for some forgotten reason, I was relieved of my Saturday evening class. During my stay in Athens, on a bulletin board at the American embassy, I had found the name of a woman who was looking for someone to help care for her dogs. I hadn't applied; I had no experience with dogs, outside of the princely Buffy. But I was curious to know what I had passed up, so, on the pretext that I was still looking for work—which I could have been, depending on the attractiveness of the arrangement, and the temperature of the house—I wrote to the dog owner, and she invited me for the weekend.

Gerta lived in Galaxidi, a handsome, once-prosperous seaside town a little more than halfway to Athens. She had fixed up an old sea captain's house for herself and her six dogs, all of which had been abandoned. The locals, she told me, had no affection for the animals; mothers told their children that if they misbehaved, a big bad dog would come and get them. From an early age, a fear of dogs was instilled in Greeks.

This didn't, I assumed, make her a popular figure in town, especially when she walked the mutts—some as big as Alsatians—through the streets every morning and evening. I accompanied her on these walks, which would be part of my duties were I to accept the job, while she talked incessantly. I assumed that another, perhaps more important, part of my job would be to listen to her. She was not uninteresting. An independent woman of very strong views, she had had a successful business career back home in Germany before her retirement to Greece. She spoke Greek fluently and knew the history of the country and of the town. When she talked about the locals, it was with a kind of anthropological interest, as if she saw them more as specimens than neighbors. She noted, for instance, that many of the men's heads were flat in the back.

Most of her attention, however, was focused on her dogs. Almost any subject would lead, eventually, back to them. She showed no interest in me, even when I dropped potentially intriguing nuggets like the fact that I had been living behind the Iron Curtain. She never asked me a question. It was all about her, her dogs, and their little world. One-sided conversations are often a symptom of living alone. Yet *I* was now living alone—really alone, for the first time in my life—and it hadn't turned me into a monologuing bore. But I had my writing for the evacuation of my thoughts. And as a journalist—I didn't dare yet call myself a writer—I was accustomed to being a listener, a role I have played, finding occasional delights in a monotony of wastelands, my entire life.

On Sunday we went to visit friends who lived in another old sea captain's house. The white interior was attractively furnished with dark antiques. We began the conversation by talking about the weather. The husband was an Anglophile Levantine; the wife, a repatriated Greek. I mentioned my surprise at how cold Greece was.

"April," the man said dramatically, "is the cruelest month."

"Chaucer," I noted, happy to be back in my own language, back in my own field.

"Eliot," he corrected me.

A weeks' old edition of the *New York Times* lay scattered on the sofa; books sat tightly bunched on shelves. Everybody lamented living among the unlettered.

"The other Sunday," said the wife, "I was sitting in the garden reading a book. And I overheard my neighbor say to her husband: 'Look at the poor woman—she has nothing to do.'"

On the walk back to her house, Gerta went on and on about her dogs. It didn't help my self-esteem to be upstaged by canines. People who only talk, and expect you only to listen, deprive you of not just your voice but also your identity. You are nothing to them but a receptacle for their words. You shrink in consequence before the verbal barrage, powerless in the dull dictatorship of the tongue.

At least that's how I felt that weekend.

Escaping to my room, I got undressed to take a much-needed bath and discovered a sore on the right side of my scrotum. It was small, like a large pimple, and, had it been anywhere else on my body, I wouldn't have given it much thought—a little thought, sure, as any low-level hypochondriac would—but it was in a tell-tale region of my anatomy. My immediate thought was of a sexually transmitted disease, even though I hadn't been having any sex. But at this point I wasn't putting anything past Greece.

Before heading to the modern bathroom with lockable doors, I used the trimmer on my shaver to cut my hair. It was the first haircut I'd given myself since Athens, and thick black curls soon littered the stone floor. I swept them up into a newspaper and stuffed it in the wastebasket.

Saying my goodbyes a few hours later, I told Gerta I'd think about the position and let her know. Then I boarded the bus back to Arta. It was heaven to be on my own again. There's nothing to make one appreciate solitude like a weekend with the self-absorbed.

Not long after departing, the bus made an impromptu stop to pick up a young couple. They captured everyone's attention, or at least all of mine, through a combination of the woman's magnetism and the man's Zorba-like exuberance. He boarded the bus as if he were embarking on a great adventure. They took a seat across the aisle from me, and we soon got to talking. Ion was English, but of Greek parents, and had recently run a Greek restaurant in London. Smiley had worked as a model in the city, after moving from her home in Rhodesia. They were now headed back to the house they had built on the island of Trizonia. It sounded like a kingdom in a Marx Brothers movie. Before they got off, they invited me for what Ion called "a bloody Greek Easter."

The traveler walks a constant line between boredom and amusement, with a sliver of danger at the edges.

I returned to my monastic existence in Arta. I wrote a letter to Gerta, declining the job because, as I told her, after only two days in her house I had started to take on the characteristics of a dog. She had, I explained, only to check the wastebasket in my room to find evidence of shedding. It was a juvenile message, but it felt very good writing it, having the last word, the writer's vindicatory consolation. I waited, feverishly, for letters from Hania, which came with an infrequency that seemed almost heartless. Didn't she know how much I was suffering? No, she didn't, because I was making light of it in my letters. And, unlike me, she still had a life; she was surrounded by friends and aunts and cousins while I had nothing to do but write. One Saturday afternoon, when I had told myself there had to be a letter, the mailman failed once again to deliver one. He was, in my mind, partly to blame. I returned to my room and flung myself on the bed, pounding the cold walls with my fists, raising myself to my knees, rapping my hands and arms against the stone as if trying somehow to climb out of my despair, my forsakenness, my wretched life in Arta. I wondered if I was going mad.

After my class I joined the evening promenade and then drowned, or at least moistened, my sorrows with a glass of ouzo.

On Sunday I walked to the stone bridge, by far the loveliest thing in Arta. It had been rebuilt on several occasions, most recently during Ottoman times, and consisted of four main arches, one significantly larger than the others. It was partly this asymmetry that gave it its beauty. Like the covered bridge back in Hunterdon County, it provided me with a place for retreat and contemplation.

I knew my stay here was temporary, but in my unhappiness the end seemed very distant. It was only March—I could still see my breath in my room—and I wouldn't be leaving until June. The town held no attraction for me. The restaurants were grim, with huge simmering pots surfaced with grease and circled by flies. Once, in one, I asked for an omelet and got a blank stare. I couldn't even find a gyro place. The baked trays coming out of the communal oven behind the church—moussaka, pastitsio, baklava—looked enticing, but they were for families, not solitary foreigners. The weight I'd lost in Poland was joined by more pounds in Greece.

The school director continued to give me the silent treatment, even when, once a week, he drove me to a nearby village to teach a class of high-school students. During the first lesson I had stopped mid-sentence to watch a shepherd lead his flock down the street. My captivation caused the students to laugh. This man who seemed to me to have stepped out of the Bible enjoyed, I was told, a rank slightly above that of the village idiot.

One girl would talk to me after class while I waited for the director to return from the café. She was intelligent and comely, even in her clunky work boots, and she asked me pointed questions about American imperialism. As startling as her fluent English, and her quietly precocious concerns, was the presence of anti-Americanism in a dusty village of shepherds.

I had a small class of teenagers in Arta as well, and on slow evenings I taught them folk songs. Ralph McTell's "Streets of London"—"How can you tell me, you're lo-one-ly?"—took on great meaning for me in Arta. But none of these young people were as bold as the village girl. After class they, like the teachers, would all head home, passing numerous cafés as they went. In Warsaw, where it would have been wonderful to talk with students and colleagues after class, there were few cozy cafés and none with outside seating in March. It seemed unfair that the absence of an outdoor culture invariably produced people—owing to an indoor life of bookish pursuits—who were the very types one wished to sit with at sidewalk cafés. Because I didn't speak Greek, I couldn't tell if the presence of such a culture had the opposite effect.

There was one teacher who was confident enough in his English to talk to me. For a holiday, he gave me the name of a former colleague who lived in Corfu, and I took a bus, and then a ferry, to what seemed like the most beautiful place in the world—ancient stone walls draped with purple bougainvillea. This is not a town, I thought; it's the backdrop for an opera.

I had tea with the retired teacher and her widowed mother in a stuffy apartment. One morning, walking past a café, I watched as a well-dressed man picked up his wicker chair and aimed it playfully at his friend. A few seconds later a female jogger ran past—not an everyday sight in 1979—and the men stared at her with a mixture of disbelief and incomprehension. On the ferry back to the mainland I kept my eyes peeled on the horizon, trying desperately to make out Albania.

The following weekend in Arta, the teacher and I went to see Paul Mazursky's *An Unmarried Woman*, which I had missed when it had come out in the States the previous spring. At the intermission, which was still observed in the town cinema, he looked at me as he never had before. "Do American women really talk like that?" he asked almost breathlessly. He was referring to the frank discussions the female characters were having about sex. I told him I was no expert, but as far as I knew, they did.

The American woman and her husband invited me to a party at their home. Everyone sat in straight-backed wooden chairs that lined the walls; I was seated between two middle-aged men. A plate of olives came around, and the man on my left poked repeatedly with his toothpick before giving up and passing the dish to me. I stabbed an olive on the first try. "Well, sure," the man said, in perfect English. "I tired it out."

In the cafés, I began having imaginary conversations with friends from the newspaper. I could go for an hour or more on a manufactured tête-à-tête, sometimes mouthing the words and even making the appropriate hand gestures. It was an absorbing, if perhaps worrisome, fantasy life for someone who found writing fiction difficult. I paid so little attention to the people around me that I gave no thought to the impression I was making.

I also started drawing a birthday card for Hania. Her birthday wasn't until April 6th, but who knew how long it would take to get there? I sketched a café—draw what you know—with a tiled roof and a tree whose branches flowed into a trellis. Then I drew six patrons sitting at three outdoor tables. On the left, Stanisław Lem sat, cross-legged, with a generic, unclothed, ancient Greek man, also cross-legged and giving the science-fiction writer a suspicious look. An Orangina-shaped bottle, with two straws sticking out of it, stood on their table. On the right, President

Jimmy Carter, in one of his trademark cardigans, sat with a pensive Valéry Giscard d'Estaing. In the middle, our man from the visa office, sweatered as usual, gave the "V" sign while putting his arm around a dark-suited man who raised his glass and looked like a young Aleksandr Solzhenitsyn. In the background stood a small waiter, in white apron and black bowtie, whose mustache and round glasses bore a striking resemblance to my own. Carved into the tree trunk were the words: HAPPY BIRTHDAY HANIA.

The card, done first in pencil, then in ink, kept me busy for days. I carried it from café to bench and back again, fine-tuning the lines, adding more detail. It was my most elaborate work as an amateur cartoonist, and, when I took it to the post office, I prayed it would find its intended recipient in *Πολωνία*.

Finally, Easter week arrived. I took the bus to Athens and found a cheap hotel near the station. It felt wonderful to be in the capital again. (Tourists who complain about the city should be sent to Arta for a month.) I reveled in the noise, the fumes, the crowds, the spring-like temperatures. I visited bookstores, ate gyros, read the *International Herald Tribune* in the lobby of the Grande Bretagne, taking my café occupations

to a higher and more luxurious level. Great hotels are like cathedrals, beautiful edifices in which you can linger, gather strength, collect your thoughts, even if you're not a member of the flock. The armchairs scattered about are more comfortable than pews and, as in church, rarely does anyone question your presence. And even if you can't afford a drink in the bar, you can always avail yourself of the elegant restrooms.

Waiting to cash travelers' checks at the American Express office across Syntagma Square, I listened to three tanned women conversing in an unrecognizable tongue.

"What language are you speaking?" I finally asked them.

"Navajo," said the tallest one.

There was something infinitely absurd and oddly appropriate about hearing an ancient language of my homeland, for the first time, in the epicenter of the classical world.

I had dinner with Christos, who was still waiting for Linda, and visited the library of the American embassy. In the stacks of periodicals I found a story by Blaine from the *Washington Post*. "I don't read you often," I wrote to him on a postcard, "but I read you widely." We were both following our separate career paths, though his was a lot more clear-cut than mine. I also found a copy of the *New Leader* that carried on its cover a beautiful black-and-white photograph of Warsaw's Old Town. Inside was an article by an American writer and editor named Abraham Brumberg who, on a recent visit to Poland, had received a proposition to become an informer. So my experience was worthy of publishing. I made a mental note to write to him when I got home.

My past weighed on me as I entered the Polish embassy to apply for a two-week tourist visa. It was a less imposing building than the consulate on Madison Avenue, which helped calm my nerves slightly. I presented my documents and the necessary headshots in the designated size and then made my way to the station.

Buses lunged with luggage on their roofs, squat women darted with wicker baskets on their arms, and the occasional man sprinted with a goat carcass slung over his shoulder. Good Friday in Athens.

My bus was delayed, so I went outside and leaned against a stone wall to savor the scene. A tall, thin boy of about fifteen approached, carrying a wooden tray piled high with *koulouria*, the dry Greek version of a sesame-seed bagel. I assumed he wanted to sell me one, but instead he placed the tray on the wall and extracted another tray from underneath it. This one he quickly filled with excess *koulouria* from the first and then lifted it to his shoulder. Before heading into the station, he told me that I was to sell the remaining *koulouria*. I didn't understand his Greek, of course, but his request was obvious. Luckily, I knew they sold for four drachmas apiece.

Business was good, and transactions went smoothly if wordlessly. I was happily transported back to my days selling scones in the food hall in London. Then a woman approached and handed me a one hundred drachma note. I gave her an I-don't-have-change-for-that shrug, but she insisted, partly because her little boy had already started nibbling on his desired *koulouri*. With no other recourse, I spoke to her in English, explaining that I didn't have change, I couldn't break her large bill, I wasn't a *koulouri* vendor by profession. Whatever astonishment she might have felt was eclipsed by anger. She ripped the *koulouri* out of her son's mouth, slammed it down on the tray, and stormed off toward the station.

It had been an extremely therapeutic moment: After months of being baffled by Greeks, I had managed to thwart one of them. When the vendor returned with his empty tray, I handed him the money and showed him the few remaining *koulouria*. He appeared pleased.

The bus finally arrived, and I took a seat in the front, where I could admire the driver's decorative taste. Just above the large windshield ran a row of portraits—reminiscent of a high-school yearbook—of the driver, men I assumed were his brothers, and Jesus Christ. Above this hung a picture of Ioannina's soccer team, an ad in English for Campari, a religious icon, several loops of worry beads, and, hanging from the rearview mirror, a raccoon tail.

Eventually we emerged from the city and began passing, every few miles, the same billboard. It advertised a popular brand of retsina and showed a seated patriarch, in white shirt and black vest, drinking a glass while self-satisfactorily turning a spit.

Soon fields stitched with olive trees and roadside gatherings of newborn lambs appeared. For the paschal season, a paschal *paysage*.

Not far from Galaxidi I changed for a local bus to Hania—the two names evoking opposite emotions. On this bus, the picture of the soccer team seemed a bit more recent. We made frequent stops, not for passengers but for sticks, which the driver collected in fields and then loaded onto the roof, while everyone waited patiently.

From the dock in Hania I could see Trizonia about a quarter of a mile out. A small, flat-roofed cluster of houses hugged the harbor. It presented a tranquil scene but still not the one people picture when they hear the words "Greek islands." A woeful fisherman appeared and took me across in his motorboat. I spotted Ion and Smiley's house sitting by itself at the top of a hill.

After a strenuous walk up a dirt path, I came to what looked like a loosely constructed shelter. Chickens strutted about the yard; a pirate flag flew from a tree.

"You're just in time for tea," said Ion, greeting me at the front door. "Did you have a nice ride?"

"We stopped for sticks," I said.

"That was probably Costas. For his Easter Sunday spit."

"He had a picture of Ioannina's soccer team."

"They all have a picture of Ioannina's soccer team. I'm thinking of doing a book on Greek bus decoration."

This, I learned during tea, was one of Ion's numerous projects. In addition to the restaurant business, he had had brief careers in screenwriting and marketing. For the moment, he was content to live off the rent he collected from the foreigners whose boats he looked after in the winter and on fees from odd construction jobs in the village.

"We haven't got electricity," he told me proudly. "There's a gas oven—no refrigerator. We use kerosene lamps in the evening. Water comes from the village. We get it from a hose I stretched all along the path. The villagers were incredulous—and some still don't believe it works."

"What do the villagers think of you?" I asked.

"They call him 'The Mad Englishman on the Hill,'" said Smiley.

"I don't talk to them much," said Ion. "There are hardly any young people here; they've all left the island. Most of the villagers are retired sailors and farmers. And what are you going to talk to them about—goats?"

On Saturday morning Ion and I rose early and, after a quick breakfast, he led me down the dirt path I had climbed on Friday. Before reaching the village, we came to a field in which men stood around a tree from whose branches three goat carcasses hung. The men walked up to the suspended bodies with knives, made long incisions, and then tugged off the coats with strong downward thrusts. Women watched from a distance while children laughed and chased each other. A red pool of blood slowly expanded beneath each carcass. I assumed this was what Ion had had in mind when he invited me for a "bloody Greek Easter."

One of the men approached Ion and led us to a neighboring field where, I soon discovered, our holiday dinner still grazed. Ion gave his approval of the animal and instructed the man to keep the head.

"You crazy, man," the Greek said laughing. "The head is the best part."

I remembered Linda telling me about her first Easter with Christos's family. A goat head sat in the middle of the table like a centerpiece, and during the soup course his mother reached over and cracked the skull open. She scooped out the brains and plopped them into everyone's broth. Then, after plucking out the eyeballs, she dropped them into Linda's bowl, as she was the honored guest. I was relieved I had no such title on Trizonia.

Ion and I went to a café to await our decapitated goat. When it arrived, Ion paid and threw it over his shoulder, just like the men I'd seen at the bus station. As we walked back to the house, I snapped a picture, one of the very few I took in those years. I'm not sure why I didn't use my camera more; it was partly my belief in the supremacy of words but also an assumption that picture taking was what tourists did, and this—Greece *and* Poland—had nothing to do with vacation. In any case, the photo shows Ion walking along a narrow dirt path high above a calm blue sea. His jeans are stuffed into his wellies, his green army jacket covers a thick sweater of thin red-and-white stripes, and a white knit cap with a white pompom envelops his head. He is unshaven and smiling while gripping the goat by its hind legs. A nearly camouflaged Yorkie walks at his feet.

Looking at that picture now, I realize the miraculous power of photography to freeze a moment and hold it forever, impervious to time and the tendentious meddling of words.

Back at the house, the vegetarian Smiley had placed a bedsheet on the ground outside. Laying the goat down, Ion took out a knife and slit open the stomach from hind legs to forelegs. Then he inserted his hand, retrieving various parts. The offal was discarded except for the liver—still warm, like a gray, offensive jelly—which Ion would later make into paté. Not a big meat eater, I have never graduated to organs, a failing that marks me as an American in most of the world.

Next, Ion stuffed the goat with oregano, bay leaves, and thyme before stitching the gut closed with a length of old wire. Then, picking up the knife again, he punctured small holes in the flesh and filled them with peeled slices of garlic. A thick branch was found and thrust in one end and out the other.

"We'd better carry him inside so the chickens don't get him," Ion said.

"But what about Smiley?" I asked.

"We'll put him in the bathtub."

After lunch I went for a nap on the couch. Awakening, I put on my glasses and, looking up to the loft where my hosts slept, glimpsed a topless Smiley sitting on the bed and putting on a sweater. Even though her back was turned towards me, and only the side of one breast was visible, it was the most erotic thing I'd seen in Greece.

A few hours later, Ion led us by flashlight down to the village. The church was packed, men standing on one side, women on the other, under two large, low-hanging chandeliers. We found places in the back, overseeing a constant trickle of parishioners heading out into the fresh air for a smoke, or a chat, before shuffling back inside. The restlessness I had to contend with in the classroom carried through to adulthood, even into church.

The priest stood hatless and with his braid untied, so salt-and-pepper streams flowed down his back. He chanted a prayer, after which the lights went out, pitching the entire sanctuary into darkness. This symbolic moment, awaiting the knock on the door behind the altar—Christ is risen!—had a riveting effect.

The priest, who had disappeared in the blackout, emerged from behind the door with a lighted candle. The congregation rushed up to him to light their own candles. When all were aglow, everyone slowly headed outside. At the stroke of midnight, a few minutes later, church bells pealed and fireworks exploded.

"*Christos Anesti!*" (Christ Is Risen!), people greeted one another. "*Alithós Anesti!*" (Truly He Is Risen!), came the ancient response, followed by kisses and a few singed lapels.

Residents from the mainland returned to their boats and glided home with candles still burning. The smooth black surface of the water mirrored the diminishing points of light.

The local taverna owner invited the three of us back to his house. We took our places around a rough wooden table in a stark white room lit by a bare electric bulb. Atop the table sat a bowl of bloodred eggs. We each took one and knocked it against our neighbor's, the cracks letting out the spirit of Christ. Then the wife served boiled goat's liver wrapped in lamb's intestine. My punishment for being an accessory to hircine murder. Smiley looked imploringly at Ion, who told her it would be rude to refuse. She picked up a gray-brown cluster and swallowed it bravely, as did I, with as little chewing as possible. The rank taste of organ curdled my mouth.

Easter Sunday was anticlimactic. The day was wet, the goat greasy. But Ion was in fine holiday spirits, and Smiley was as well, the two of them happy to have a guest at their idyll, while I felt like a soldier on furlough.

I got the bus back to Arta and my cold room, my bench in the sun, my tsking students, my nocturnal cafés. I wrote to Hania about my colorful Greek Easter and picked a night, at the end of April, and an exact time, at which we should both look up at the moon—momentarily forgetting, in the blueness of Greece, that Poland was often overcast. I didn't mention my sore. A letter arrived from my mother telling me about the accident at Three Mile Island and how she had calmly driven out to see her mother in nearby Mechanicsburg. My brother Jim, in a rare epistolary moment, wrote to tell me that I had become an uncle. He wondered what the

Greeks thought of the newly signed Egypt-Israel peace treaty, and of course I had no idea. I made a note to ask little Socrates.

One Friday I took the bus back to Athens in hopes of picking up my visa. Outside the embassy, I ran into a young American who had also been living in Poland. In fact, he had taken the same series of trains to Athens that I had. I tried to exchange war stories with him, but he wasn't into it. The five days of sitting in cold compartments with exotic strangers through communist Europe had made little impression on him. He had ridden those trains as nonchalantly as my mother had driven into a potential nuclear disaster, and I admired his poise, his unruffled man-of-the-world demeanor. He seemed the ideal of the accomplished traveler, cool and unfazed, while I was a fearful, pathetic amateur. He made me question my chosen profession.

Later, though, it occurred to me that it was the well-adjusted ease with which he moved through the world, and his congenital oblivious-ness to it, that made his journeys such nonevents and his ability to write about them—had he been asked to, since he clearly would never have felt the need—almost nonexistent. By contrast, my sensitivity to my surroundings, my capacity for awe, my live-wire imagination (at least in the realm of personal calamity) all caused me to experience life more as a drama—sometimes comic—than as a breeze and qualified me to be a chronicler of my travels. The beauty of being impressionable, a word usually used pejoratively, is that it leaves you with a lot of impressions. It is often said that the worst trips make the best stories, and a corollary to that is that bad travelers often make good travel writers—and not just because we're constitutionally prone to having bad trips.

But in the immediate aftermath of my encounter with the dauntless, or perhaps just callous, American, my self-confidence dropped to a new low. I entered the embassy half-believing that the Poles should refuse to give me a visa based on my overall unworthiness. Who was I trying to kid—living abroad, riding trains, crossing borders, petitioning embassies? Forget the fact that I had already been implicated in a standoff with visa officials in Warsaw, a seemingly irreversible cause for blacklisting; I would be denied a visa because I didn't deserve one.

A combination of bureaucratic intimidation—communist at that—and months-long loneliness made me a quivering supplicant that morning.

Inside, my visa was waiting. No questions were asked.

I walked out quickly into the Athenian sunshine, stunned—even in that pre-internet age—by the woeful, if personally beneficial, lack of communication between government offices. It meant that my plan of coming to Greece for the winter, however disappointing in nature, had been fundamentally sound. Yes, I had known suffering, but in my hand I held the visa to Poland, to Hania, that made it all worthwhile.

On the bus back to Arta I experienced a sore throat that developed, during the course of the trip, into a swiftly running nose, achiness, and fever. By the time we pulled into town, everything was closed, even the cafés. I had nothing to do but return to the room where my breath was still visible. Without undressing, I got under the chilled covers and pulled them as far over my head as I could without suffocating.

When I awoke the next morning, my cold was gone. Another unexpected gift. Without medication—I didn't even have aspirin—in an unhealthily damp and frigid room, I had defeated a cold that, normally, would have nagged me for a week. Still wearing the clothes I had worn in Athens, I sat on my bench and wrote a letter to Hania about my impending arrival in Warsaw.

Spring finally came to Arta. One Saturday the plumber and his wife invited me to a house they rented in a nearby seaside town. Everyone sat in the shade of the courtyard except me, the heat-seeking foreigner. A neighbor offered me a seat in the shadows, which I declined. The plumber gave the man an explanation, the only word of which I understood was "Polonia."

It wasn't just warmth I was seeking. The ideal then was "tall, dark, and handsome"—and probably still is, though it's not stated in an age when everyone is supposed to cherish diversity. Feeling myself lacking in two of those attributes, I worked hard at the one I had some control over.

We ate Greek salad, snowcapped with feta, and fried calamari drizzled with lemon juice. In the evening, a tape player was plugged in and everyone danced. Mothers, fathers, teenagers, grandparents, children—including the plumber's two little girls. A long line formed and bent according to space. I loved the inclusiveness: Through linked arms or held hands, the child and the elder, the expert and the novice, the local and the foreigner, moved as one.

During my final days in Arta I met a young man who had recently opened a competing English school in town, and on my last night he invited me to dinner. At a restaurant on a street I had somehow missed, I ate the best meal I had had in Arta: roasted chicken with roasted potatoes.

The man told me that he had worked at my school a few years earlier. At the end of the semester, the director informed him that he needed to pass all of his students. That was the rule: Every student, regardless of performance or ability, passed to the next level. When he asked why, the director explained to him that, in the past, whenever students were held back, they dropped out, and the school lost money. So, year after year now, the charade had continued: Students enrolled, most progressed without learning, and their parents, under the impression their children were becoming fluent, happily paid their annual tuition.

I was glad I had not been involved in grading. I wondered if I was supposed to have been, and it was just one of many things the director had failed to communicate to me. Of course, whatever I would have done as a departing teacher he would have negated as soon as I had left.

We lingered over our sidewalk meal. I was extremely grateful for the company, the conversation—the man's English was excellent—and the insight into the real workings of the place. And, of course, the food. But I couldn't help but note the extremely bad timing. It was my introduction to an unfortunate travel phenomenon, one that has plagued me occasionally as a travel writer, and that is that sometimes the moments of interest and promise one hopes for on a trip arrive infuriatingly on the last night.

It was late when we got up from the table. I wished the young man good luck with his school and headed back to my finally comfortable room.

In the morning I said goodbye to the family—the two little girls, I noticed happily, looked a little sad—and lugged my heavy suitcase down now-familiar streets. Then I boarded for the last time the bus to Athens.

That evening I had dinner with Christos and Linda, who had arrived a few weeks earlier. In an excellent mood, and a rare state of cleanliness, I had put on my favorite shirt: a colorful and expensive plaid button-down from Macy's that Hania had convinced me to buy on a trip to Manhattan. Christos maligned it, saying no man in Greece would wear such a shirt. It was true, I suddenly realized, that adult Greeks rarely wore plaid and shied away from colors, except navy blue.

Linda told of a recent trip to the beach with Christos's family. One of the cousins, a woman about our age, frankly sized her up as soon as she stripped down to her bikini.

"I liked that," said Linda, "that she was up front about it. Americans would scrutinize you, but from a distance, so no one would know."

After telling of their life, they offered advice for mine, specifically my upcoming reunion with Hania.

"Be tough," said Christos.

"Be understanding," said Linda.

The following day I headed out to the airport to find a hotel for the night. The mere sight of the airplanes excited me. I had flown a number of times, of course, but never had my joy at what I was traveling to been matched by my delight at what I was fleeing from. Four months in Arta, helped by a recent letter from Hania, had created the perfect traveler's bliss.

Sleep proved difficult because of my pleasantly agitated state—and a mosquito that delivered a series of parting shots.

CHAPTER 4

Euro-Jersey Limbo

SITTING BESIDE HANIA IN THE BACKSEAT OF THE TAXI—MUTE AS always, as Hania had taught me that the driver would up the fare if he knew he was carrying a foreigner—I had a strange sensation: that I was back in the West, not back behind the Iron Curtain. The streets of Warsaw looked much more contemporary than those of Arta and even, in some ways, Athens. This was in large part due to the women—especially the young—striding confidently in their summer dresses. It took Greece to show me how European Poland was.

My time in the two countries had revealed an irreconcilable conflict in my nature: I am attracted to southern climates and northern mentalities.

Adding to my pleasure at seeing Warsaw again was the understated holiday atmosphere, for my return coincided with that of the new pope, who was making his first trip back to his homeland since becoming pontiff the previous fall. This historic visit was bringing international media to Poland—for the first time in years—and it made my gift of a visa seem all the more remarkable.

The taxi dropped us on a leafy street of old apartment buildings. A few months earlier, Hania had moved to a gracious neighborhood a short tram ride away from Plac Zbawiciela. Sharing the ground-floor apartment with her were her faux-aunt Marylka, a stocky, white-haired woman with lively brown eyes; Marylka's daughter, Elżbieta; and an elderly boxer named Babichon.

We took my large leather suitcase to the high-ceilinged bedroom and dropped it next to the double bed. I pulled Hania on top of it and she got up laughing, noting the presence of her aunt in the kitchen.

The pope wasn't arriving for a couple more days, so in the meantime I reacquainted myself with the city. I visited the English Language College and found Jolanta, just back from her year at the University of Michigan, in her family's apartment. Her suitcase lay open on the floor, one side of it filled with Reese's Peanut Butter Cups. I made an appointment with a doctor who identified my sore as a boil and asked if I'd been bathing regularly. I went to the Thomas Cook office on Nowy Świat and reserved a seat on the overnight train to Paris. I played tennis with Peter and Jeanne on red-clay courts on the other side of the Vistula—it was a new and wonderful feeling to sweat—and, afterward, we went to a café and downed glasses of unsatisfying lukewarm juice. The Poles were unaccustomed to the heat and, standing in lines for mineral water—my beloved Mazowszanka—they looked as miserable as they had in their winter queues. They seemed a people born to suffering. It came through in their inflections—incapable of understanding their words, I was attuned to how they were delivered—which, even in many of the young, especially the females, carried a note of almost caricatured melancholy.

Marylka's speech was often woeful; she was of the generation that had survived the war and postwar Stalinism, a period Americans tend to overlook, at least its Eastern European manifestation. Because for so long we viewed this part of the world as one faceless, forgettable whole, many Americans are oblivious to the cruel irony that Poland lost the war despite fighting, valiantly, on the winning side in it. Not far from her new apartment stood the prison where Hania was born.

I spoke French with Marylka, who had learned it at Szymanów, the Catholic boarding school that Hania and her mother had both attended. Relieving her melancholy was a rebellious spirit and a keen sense of the absurd, which was as important as patience to survival in Poland.

On the morning of June 2nd, we walked a few blocks to join the crowds lining Wawelska Street to greet the pope's motorcade. Yellow-and-white papal flags waved in unison with white-and-red Polish flags. When the popemobile finally appeared, the man inside received the welcome of a returning hero. With the love and admiration, though, came a beseeching hopefulness. The pope was the ultimate immigrant in a country with a long history of producing them, and it was clear that Poles looked to him as someone who could improve their lot—perhaps, as the Vicar of Christ, perform miracles.

Peter and Jeanne came by the apartment and presented us with a children's book, *Larry the Lizard*, that Peter had written and Jeanne had illustrated. Then the four of us headed downtown. (Marylka was staying home to watch the mass that the government, in hopes of reducing attendance, had decided to televise.) Hania scored four tickets from a scalper on Marszałkowska, astounding us all with her street savvy, and we walked

briskly through a city transformed. There was a clear but covert sense of expectation, an unfamiliar electricity before an unprecedented event, as citizens headed purposefully, reverently, defiantly to an open-air mass to hear their pope.

A thirty-foot tall cross, draped in two red stoles, lorded over Plac Zwycięstwa (Victory Square). Behind it, rubberneckers bunched on the balconies of the Hotel Europejski. Hania led us through the tightly packed crowd—so much for the lure of televised history—and found us a space in the shade of the Hotel Victoria, next to a group of women from Łowicz in traditional dress, their skirts billowing around them like colorfully striped pinwheels.

Mass was said, with a homily whose every word was savored, or loathed, depending on where it was being heard. You had the feeling that the whole world—not just Poland, Washington, and Moscow—was listening. I understood almost none of the sentences but sensed by watching the people—one man standing alone, his hands clasped behind his back, his head bowed in a look of intense concentration—that they were having a profound effect, as was the emboldening spectacle of thousands of people gathered together on the main square of the capital professing beliefs at odds with their government's.

We left early and walked down Nowy Świat. There were almost no cars, and the few other pedestrians walked as if in a daze. Adding to the unreality of the moment, especially for Hania, were the speakers on lampposts broadcasting the mass. Now I wasn't the only one experiencing a feeling of geographical confusion.

A few days later Hania and I took a tram to a less lovely neighborhood, where she was looking to buy an apartment. I had hoped she'd been thinking of moving to the States. Be tough, I told myself, be understanding.

I boarded the train to Paris—Hania would join me in a week in Strasbourg—and, getting off at Gare de l'Est, I was greeted by Dan and Jane holding a bright red KitKat bar. I had written to them of my arrival, though it seems almost inconceivable today that words scrawled on paper that was inserted into a box and then carried across borders could have been received in time for the message they conveyed to be acted upon.

Yet I fully expected to see my friends at the station—Jane was spending the summer in Paris—just not with my favorite dessert from the *Trenton Times* lunchroom.

This visit to Paris provided the first indication of my obsession with Poland. I found a Polish film, Wajda's *Bez Znieczulenia* (*Without Anesthesia*), and went to see it with Dan and Jane, though I arrived two hours early so I could watch it twice. It told the story of a Polish foreign correspondent who returns home to criticism that he avoids writing about the problems in his own country. We didn't know at the time that the character was loosely based on Kapuściński. The title of the film was taken from a scene in which the man refuses anesthesia for the extraction of a tooth, one of a number of difficult moments that caused Dan to remark, after I'd told him I'd sat through the earlier showing, on my dedication to all things Polish.

I discovered a Polish bookstore on Île Saint-Louis, run by a soft-spoken man with a high forehead and casually elegant attire. Though Polish, he seemed perfectly at home on that exclusive island in the Seine, which made me think of something my Polish teacher's husband had told me: that Polish jokes didn't exist in Europe because the Poles who had immigrated to France and England were often the well-educated—writers, artists, intellectuals—who harbored hopes of returning one day. By contrast, it was the peasants and workers who had come to start new lives in the States. There have been obvious exceptions, like the miners in the north of France, who gave birth to the term "*les bandits polonais*," and the more recent creation of the European Union: "the Polish plumber."

A larger bookstore—Librairie Polonaise-Księgarnia Polska—stood on Boulevard Saint-Germain. A store devoted to Poland on one of Paris's most fashionable streets seemed like a justification of my new passion. One day, going through a pile of Polish posters—a few of which I had admired on kiosks and hoardings in Warsaw—I struck up a conversation with one of the shop assistants, a young man by the name of Stefan. I told him, in French, of my time in Poland, especially about the recent visit of the pope. He invited me out to meet his émigré friends, introducing me to the bond that is quickly established between Poles and people with an interest and a stake in their experience. This bond, I learned later, is even stronger when you speak their language, an ability that automatically puts you into a select group and makes you almost like a member of the family.

I wasn't researching a story, but I was demonstrating an ability to talk to strangers and establish friendships that would serve me well as a travel writer.

After Dan left, Jane invited me to a soiree with some of her fellow grad students from Princeton. A quiet young man named Ulrich told me of a scene he had once witnessed in East Germany, near the Polish border. A table of Polish customers were exhibiting extreme deference to their German waiter. I told him that Poles put great value on politeness. But their behavior toward the German was so perfectly, exquisitely mannered that it revealed, he said, their inner contempt.

Ulrich showed the cartoons he'd done during his year abroad. The drawings were clean, spare, angular—a style the *New Yorker* wouldn't publish for another decade—and the humor was sharp. One showed a man smoking alone in a room, above the caption: "Cigarette after masturbation." They made my drawings look like the work of a child.

We headed out just as the sun was coming up. A milkman passed by, and the Princetonians engaged him in rapid-fire repartee, only half of which I caught. The world, at least the Anglo-Gallic one, lay at their feet.

I took the train to Strasbourg, where I met Hania at the station. "Come up here and help me with my bags," she ordered from the steps of her *wagon*, in a tone I'd heard before, which she thought playful and I found bossy, especially since she employed it almost exclusively in public. It was part of her lack of sentimentality and, perhaps, her need to demonstrate power and independence, even when requesting male assistance. Happily, it was always short-lived and, if I brought it up, dismissed as nothing. If I mentioned it stung, I was told I was overly sensitive.

We spent some time on the farm. Free room and board was definitely a draw, but I was also eager to show—and Hania was amused to see—my rustic side. She bonded with Lulu and the cats in the courtyard and saw what I found so appealing in Dany. One day in the drying shed—he had started growing tobacco—he hummed "The Internationale," a tune I didn't recognize. When I was let in on the joke, it reinforced the idea that they were Europeans—with a shared past—and I wasn't. Deep in the Alsatian countryside I was the rube, the unworldly American.

I had long had an inferiority complex with regard to Europeans. It came from early contact with students who would spend a week in our home as part of a Rotary Club exchange. They were understandably more mature than I was, but I sensed that would have been the case even if we'd been the same age. They came from ancient, established cultures, countries with centuries of learning and conflict, and they carried that experience in their sentences and their behavior, which came across as more measured, less spontaneous than Americans'. They seemed to be always thinking of the consequences—though some of this was the result

of speaking a foreign language in a foreign land—while we were much more focused on the moment.

I believed that everything European was superior, more refined and sophisticated than its American counterpart: education, culture, fashion, cuisine, even sports—the elegant dance of soccer versus the violent collisions of football. When I eventually traveled to Europe I tried to adopt its ways and become as much like Europeans as possible—"as much like Oxford as monkeys can make it" were Bertrand Russell's words on first seeing Princeton. And, deprived of a rigorous, classical education, I *was* a bit simian. I knew I could never speak French like a Parisian, or even English like an Oxbridge grad, though I sometimes affected British-isms—"she's in hospital"—I had picked up from reading Waugh. (I never, however, called a radio "the wireless.") My speech had already changed as a result of adding a new language, teaching my own, and never hearing the words "I love your American accent." Mine had left New Jersey behind and landed somewhere in the middle of the Atlantic. On my letters home I wrote the dates at the top with the month in the middle, arguably because I liked the symmetry—numeral, word, numeral—but really because it made me appear Continental. Though I drew the line at smoking, and eating organs.

In Europe I saw the totems of the history that had shaped the inhabitants, and I heard from them how childish—at best, how innocent—they found us. Intellectually, artistically, romantically, we seemed to be playing in a lower division. It is difficult for any foreigner to feel superior in France—even, as I've suggested, on a farm. In Poland, my inferiority complex was less pronounced because of the grand national one; the very thing that made me insecure in Western Europe—my irrevocable Americanness—transformed me into a kind of star in Poland.

But Poland brought into even sharper focus my homegrown optimism, a sign, for war-hardened Europeans, of immaturity. I had watched a lot of European films—part of the reason for my moving to France and learning French had been my fascination with François Truffaut—so I knew that Europeans were not given to happy endings. I just hadn't thought this through enough to decipher the reason: They were fatalists—or, if you prefer, realists. The happy ending was as aesthetically repugnant to most directors on the Continent as it was personally implausible to most moviegoers. How could a country like Poland, stuck behind the Iron Curtain after fighting with the Allies—to cite just one of many national indignities—believe in joyful, or even just, conclusions? I, on the other hand, was not only an American; I was one of America's privileged youth, a baby boomer with a college education and well-to-do parents. For me, the idea that everything would turn out okay was almost an inalienable article of faith. In most of Europe this stamped me as naive; in Poland it made me positively exotic.

Hania and I eventually took the train to Paris. I knew the city fairly well—having spent a month there, between Provence and Alsace, three years earlier—and I spoke the language better than she did, which gave me some degree of confidence. Walking one day down an elegant boulevard, Hania expressed quiet resentment that Paris had been spared in the war while her hometown, which had taken a less laissez-faire approach to occupation, had been destroyed.

I introduced her to my new Polish friends, who were from the same social milieu as she. Was it socialism that, in supposedly eradicating the class system, had made it stronger, or was its strength just a byproduct of its being Polish? One of the group, a kindly, bearded man named Zbyszek, invited us to stay in his house in Argenteuil while his parents were away on vacation. While clearly not French, we nevertheless became beneficiaries of the generous Gallic attitude toward time off, which, needless to say, didn't extend to farmers.

Stefan sometimes got us into the university cafeteria, which brought back some of my worst memories of Aix, while Hania took a certain delight in discovering that her university's food was better—or at least not quite as repulsive—as that of a school in the gastronomic heart of Europe. I preferred, of course, the gyros in the Latin Quarter, which, with culinary pridefulness, were made with baguettes. Crêpes—*jambon, oeuf, fromage*—constituted a cheap and delicious meal, though the square buckwheat pancake set down on its bare white plate never looked very appetizing—resembling a thinly stuffed slate—until it bled its orangey-yellow yolk. I loved the Breton cider so much I would bring bottles of it home from the *supermarché*. Some evenings we'd go with

Zbyszek to Au Trappiste near Place du Châtelet, where the mussels came in bowls with crusty bread for dipping, and our beer—Mort Subite— arrived in stemmed glasses that were shaped like chalices. I had never tasted beer like this—European brewing superiority—and I had never sipped Budweiser from quasi-religious goblets. Paris, after the deprivations of Warsaw and Arta, truly did feel like a feast.

When we needed money we walked to the American Express office on the Right Bank, where I made withdrawals on the card my parents had given me. Hania called these excursions "trips to see our uncle," though I felt a tremendous uneasiness about them. One day we had an argument; I told her we had to better manage our money—which was my money, as the *złoty* had no value outside Poland—and thus exhibited my unattractive Protestant side. She had more of a (Catholic?) carpe diem philosophy, which was not hard to understand considering the fact that she found herself, for a short while, in the West. Why not enjoy it? In addition to eating, she did a little shopping.

We eventually made our way back to the farm and then, a few days later, traveled to Strasbourg, where I saw Hania off on the train. Whatever disagreement we had had was well behind us, though I still wasn't sure I'd see her again. What if, the thought occurred to me, she had just been using me for a spree in Paris? Returning to Kutzenhausen, I walked into the farmhouse and ran into the grandmother, who spoke only Alsatian and German. "*Elle est partie*," she said, and let out a coarse laugh, more at her unexpected French, I think, than at my unhappy fate.

I returned to Paris to get my plane and, on my last evening, went to Shakespeare and Company. George Whitman was eating his supper by the register, picking at the delicate bones of a fish. Three young men entered, each burdened with a backpack, and asked if they could sleep upstairs. He informed them there were no beds left. But, they said, someone in Boston had told them they would be able to stay at the store. "Boston?" Whitman asked. Yes, they said, they were students at Harvard. "Well, I can't turn away Harvard men," the wisp of a bookseller said and, with a small piece of fish stuck in his goatee, he gruffly told them to climb the stairs.

I walked out empty-handed but with my head filled. I'd been away from my books for a long time. I strolled along the Seine in a quietly charged state, contemplating the future with a dreamy resolve. The setting sun was giving the gray city a rare moment of color, and it seemed significant that the drama was enfolding in the west. I leaned against a railing and thought of all the American writers I loved and how they—even the undervalued humorists—had created a worthy national literature. My eagerness to return home, to dive even deeper into that literature, perhaps one day to contribute to it, grew in the gloaming. My American optimism had me thinking optimistically about America. I experienced a kind of epiphany, and it was all tied up with homecoming—not Joycean "exile," though exile had preceded it. Years later I would read a remembrance by Saul Bellow of his time in Paris and how he had gotten the idea to start writing *The Adventures of Augie March* by walking the streets and looking at the rags in the gutters directing the flow of water.

For the first time in a long while I was thinking more of my career than I was of Hania.

On an overcast afternoon in November, eating lunch with my father at the Pancake House on Route 22, I gazed at the customers perched on stools—somber, overweight men in shapeless coats—and the washed-out waitresses delivering coffee and thought: Eastern Europe has nothing on Phillipsburg.

But at home there was heat and my mother's cooking, though—having gotten used to meager portions—I often failed to finish dinner. Even after my stomach returned to its normal size, I insisted on eating my salad after the main course, *à la française*. My parents found this more amusing than pretentious. Their delight at having me back kept them from drilling me about my plans. It helped that I had two older brothers; the baby is often given a pass. They never tried to deter me from majoring in English (while expressing reservations) or going to France in order to learn French (ditto); even my decision to quit the *Trenton Times* and move to Poland to pursue a woman was accepted, if not approved of. They knew that I was different from my brothers, and they had agreed to live with it and, to my eternal gratitude, lend the occasional financial support.

In my father's law office on South Main Street, sitting with his secretary out in the front room, I typed up "The Iron Curtain Local." At the end of the story I suggested, facetiously, that the train would grow in popularity as the *Orient Express* declined because Paris, where that train originated, had lost its appeal to young American literati while Warsaw, with its low cost of living and harsh winters conducive to writing, would now attract them. Unknowingly, I was describing Prague—minus the sex—two decades hence.

In addition to some humorous pieces, I started work on a novel. In college I had taken a short story writing course, the only writing course I ever took. There were not a lot in those days; most writers were autodidacts, taking their lessons from the pages of their predecessors. The story I wanted to tell was of a young American man (surprise) who finds himself in an era of increased ethnic identity (the miniseries *Roots* had appeared two years earlier) after having grown up in a family with little sense of its past. My ancestors were German, Irish, Welsh, with perhaps some Dutch mixed in—both the multiplicity and uncertainty being fairly standard among many of my friends. I would attend ethnic festivals—like the Feast of Lights in Chambersburg—and feel envy for people, in this case Italians, who had a clear sense of who they were. It would have helped me, I imagined, to feel more European if my origins had not been spread all over the map, in some cases with uncertainty; if instead I had one country and one culture to lay claim to. My protagonist—can you guess?—meets a young Polish woman and follows her to Warsaw, where he finds an ethnicity to wrap himself in. I sent the first two chapters to an editor at Random House by the name of Klara Glowczewska.

But I hadn't forgotten my epiphany by the Seine. In a used bookstore in West Philadelphia, just off the Penn campus, I picked up two volumes of H. L. Mencken's *The American Language*. Though it is often used as a reference book, I read it cover to cover, pausing occasionally to draw caricatures in the margins. I loved, not surprisingly, when he wrote of all the loan words from other languages, and I would occasionally entertain my parents at dinner with some of the more intriguing ones. It seemed the least I could do. The scope of his research and the depth of his knowledge were incredibly impressive; he seemed more like a scholar than a journalist. But he was very much in the American grain—a self-made man of letters—and therefore was the perfect model for an embryonic writer who had been bumbling around Europe. That he had lived not in New York but in Baltimore added to his appeal. I had gone through an Edmund Wilson phase my first year out of college, when I lived in DC, reading *Classics and Commercials* and *The Bit between My Teeth* as a kind of substitute for graduate school and an antidote to the engineering reports I was proofreading for the Navy department. But for all his erudition, quite possibly because of it, Bunny was not especially funny. While not strictly a humorist, Mencken used humor as a ploy to keep readers engaged while teaching them things or, in the essays, spouting his opinions. He educated and entertained simultaneously, a trick that I hoped more reading of him could teach me. For Christmas I asked my brother Bill for *A Mencken Chrestomathy*, writing the title down because I had no idea how the last word was pronounced.

I had also fallen hard for Nabokov and started savoring his novels, going all the way back to *Mary*. Reading him in Phillipsburg was, I suspect, almost as exotic as reading him in Tehran, a city that was now in the news because of the fifty-two Americans who had been taken hostage there on November 4th. The revolution I had heard about in Athens had come home to haunt me.

One day my father interrupted my reading to show me the recent American Express bill. He wondered why it was so high. I confessed that I had taken money out of the Paris office a few times. A few times? Well, from time to time. He asked what I had spent it on. I mentioned that we had stayed longer in the city than we had planned, and expenses had added up. I took full blame; he didn't need more reasons to disapprove of Hania. Throughout the conversation, my father wore an expression I remembered from the days when I would bring home mediocre report cards—one less of anger than of disappointment—and it transported me out of the world of Russian émigrés and back into high school.

Just before Christmas I made a trip to Philadelphia with my new résumé. It barely filled one page. Miscellaneous things—like my fluency in French—were listed under the heading "Furbelow." The word was guaranteed to irk most editors, especially at newspapers, though in my naivety I imagined it would impress them.

I stopped at the offices of *Philadelphia* magazine, then one of the best city magazines in the country. The editor, Alan Halpern, is sometimes credited with inventing the city magazine. Even though I'd arrived without an appointment, Halpern graciously invited me into his office, where we were soon joined by his assistant editor, a tall, patrician-looking woman wearing large glasses and a white turtleneck sweater. There were no writing positions available, I was told, but the two of them chatted with me as if they had nothing to do that afternoon. Through the window we could see snow starting to fall. I mentioned that I had been teaching English in Poland, and Halpern picked up a book a publisher had just sent him, *Bitter Glory* by Richard M. Watt, and told me I could have it. It was a history of Poland between the two World Wars. We continued talking as the snowflakes thickened. The conditions outside heightened the coziness of the room, with its book-filled wall, its cluttered editor's desk, its two main occupants showing unnecessary interest in a flighty young man. (I couldn't call myself a writer; I had, at that point, spent almost as much time as a teacher as I had as a journalist.) There was no pressure to go, and I desperately wanted to stay, perhaps to write stories of the city without ever leaving that office. When I finally did depart, it was without a job but with a new book—I was still avidly collecting them—and an experience of editorial kindness that, I would learn, was extraordinarily rare.

At Christmas I played over and over again an album of carols (*kolędy*) I had brought back from Poland. To my mother, it was a sign of how much I was missing Hania, but I loved the songs for more than their associations. Then I rang in the New Year—the 1980s—at a party at my Polish teacher's house.

A few weeks later I drove to Washington, staying with my friend Jack, a librarian who lived in a large apartment building off Connecticut Avenue. I retain an image of Jack, a few weeks before I left the city for my year in France, arriving at our public tennis courts cradling a copy of Cyril Connolly's *The Evening Colonnade.*

I had written to Abraham Brumberg about my experience with the visa officials in Warsaw, and he had replied that if I ever found myself in Washington I should stop by and see him. Brumberg lived in a modest, comfortable house in Chevy Chase. He had been the editor of the scholarly journal *Problems of Communism*, and his wife, who served the dinner, was a professor of Russian literature at Howard University. She was quite a bit younger than he was and possessed a vaguely Pre-Raphaelite air. They were both intellectuals. Over dinner I told my informer story, and Brumberg asked if I had a picture of my girlfriend. Something in the way he said it—prompted no doubt by the quality of my discourse up to that point—suggested that he saw me as someone more focused on the problems of romance than those of communism. Though in my particular

97

case, they were somewhat entwined. I dug into my wallet for a photo; it was a student portrait that showed Hania looking serious and assured.

"She's very beautiful," Brumberg said admiringly. Then he added, as if she were one of the hostages in Iran, "You must get her out of there."

Shortly after, he excused himself, saying he had work to do. His wife and I chatted for a while, perhaps about Nabokov. Not that I would have had anything interesting to say to her about him. When Brumberg died in 2008, the obituary in the *Washington Post* noted that his home in Chevy Chase "became known as a lively meeting place for communist dissidents, scholars, journalists and emigres." But for that evening, it was just the indulgent professor and me.

Back home in New Jersey, I fell into a writing project. Dan had left *Horizon*—after the owner announced he was moving the offices to Tuscaloosa, Alabama—and become the managing editor of *New Jersey Monthly*. The magazine, with financial backing from United Jersey Banks, had decided to put out a coffee-table book about the state that would be a collaboration of writers and photographers. Titled *New Jersey: Unexpected Pleasures*, it would consist of about a dozen chapters. I chose the ones that seemed the most like travel or feature stories: "Bridges," "Meadowlands," "Cranberries," "Glass & China," "Malls," and "Paterson." Unhappy memories—personal and professional—made me pass on "Atlantic City," even though it was going to be primarily a photo essay—appropriately, I thought, in black and white.

I moved into Sally and Sam's big Trenton house to be centrally located. Buffy was as sweet as always, a well-loved but not obsessed-over pet. Celestine was still at the paper, which puzzled everyone since she was family friends with Katherine Graham. I spent a long Saturday morning at her house telling stories of life behind the Iron Curtain that still rattled me and would, I assumed, have meaning for her—not knowing at the time that she would eventually become the Moscow bureau chief for the *New York Times*. For Easter, Sally bought a round paschal loaf from Peoples Bakery in Chambersburg, and I drove the river road home for my mother's baked ham.

I was reading *Lolita* by the time I got to "Malls." "Ah, yes, the Mall," I wrote, "whose very name becomes a sound of delectation and bulk:

'Mmm—all!'" A former English major at Yale, Dan recognized the influ-ence. While I was writing about Paterson, Zbyszek arrived for a visit from Paris. I told him the city had once been called "The Lyons of the New World," and he immediately assumed it was because of its food. No, I told him, silk.

The research was extremely enjoyable. I loved driving around the state, talking to people, learning new things, resuming my old routine. Crossing the Pulaski Skyway was not great fun—it took me a while just to find the entrance—though it allowed me to slip something Polish into almost every chapter.

But I found the writing difficult. For the longer chapters—"Cran-berries," "Glass & China," "Meadowlands," "Paterson"—I had the benefit of interviews, but there were the usual problems of organization. For the shorter essays—"Malls," "Bridges"—I had to rely on imagination and style. The latter contained only one quote—from Benjamin Franklin—and very little information; it was primarily a prose poem that extolled the variety, and sometimes played on the names, of the state's many bridges. It helped that I had worked on five of them and had grown to love them more as structures than as metaphors.

To my delight, Dan liked every chapter. He edited each lightly, not-ing, correctly, that I had a predilection for the unexpected word, which could be rewarding but could also sound forced and could sometimes be inaccurate. Like all great editors, he had an infallible eye for what worked and what didn't, and, unlike a lot of them, he had the gentlest, most considerate and well-reasoned way of telling you when something fell into the latter camp.

When all my chapters were written and edited I drove over to Easton—crossing the free bridge—and stretched out on a bench on the campus of Lafayette College, my father's alma mater. My work on the book had more or less coincided with the spring semester. Looking up at the clouds, I felt a tremendous sense of relief—it's done!—and a smaller one of accomplishment. I was pleased with the finished product, but it was hard to take great pride in the fact that I was one of three writers of a coffee-table book that was going to be published by a bank. It was not how I had imagined my first book. I wasn't an author; I was a coauthor.

And I wasn't going to be in bookstores; I was going to be in corporate lobbies. My words would not be bought but gifted.

I now spent a lot of time driving back and forth on the green river road, my favorite in New Jersey, even though it had not been deemed an "unexpected pleasure"—an omission that allowed it to remain in my mind as my personal artery. Route 32 on the Pennsylvania side was even prettier than Route 29—more undulating and winding—but I felt an allegiance to my home state, particularly now, having half-written a book about it.

One evening I took my unsung road down to Lambertville, then headed east toward Princeton, for a dinner party at Mark and Sandy's. They had sold the house in New Hope—where my love life was born—and purchased one on a secluded, wooded lot a short drive from town. During the meal, Mark and Celestine's boyfriend, Peter, who for the last few years had been working on a book, talked enthusiastically about V. S. Naipaul, especially his novel *Guerillas*. They spoke with the kind of intensity that made you feel not only that you needed to read the book as soon as possible but also that you were deficient for not having done so already. The fact that you had made it through *A House for Mr. Biswas*, a much longer novel, scored you no points; it didn't carry the political clout that this work did.

Mark announced that he had accepted a job with *Businessweek*; the title, he said, was Paris correspondent; he and Sandy would be moving in a few weeks. They wondered if I'd be interested in house-sitting for the summer. There were no pets or plants to care for; my only responsibility would be to not burn the place down.

I moved to Princeton and began the kind of summer that would have been idyllic—halcyon, to use one of Mark's least favorite words—if not for my sense of aimlessness. And the absence of Hania. And the growing pile of rejections. Like a bookkeeper, I recorded all my submissions and rejections on a yellow legal pad. The list had grown to several pages and made one female friend, who saw it sitting on the kitchen table one day, moan with sympathy. Or perhaps it was shock at my masochistic nature.

All of my friends (except Peter) had jobs—as was expected in America—and the journalists (with the exception of Celestine) were making significant advances in their careers. Me, I was spending my days at the public swimming pool, where I made feeble attempts to study Polish; the campus tennis courts, where I played with an ambidextrous graduate student in anthropology who lessened a little my shame at being unemployed; and Firestone Library, where I lost myself in Nabokov's entertaining study of Gogol and, like a grad student, or the master himself, copied felicitous passages onto three-by-five notecards. He taught me a new word—*poshlust*, meaning "philistine vulgarity" and used by Vlad to damn earnest, often well-received novels—that I carried around that summer the way in high school, after reading *The Catcher in the Rye*, I had labeled pretty much everything "phony." It is not surprising that, in my feeling of inferiority vis-à-vis Europeans, I had latched on to, literarily, two men who wrote from positions of almost insufferable superiority: Evelyn Waugh and Vladimir Nabokov. It helped of course that they were both brilliant stylists. I wondered—but not too much—about the connection.

Naipaul was soon to join them. I read, obligingly, *Guerillas*, and thought the story a little contrived, even though, as Mark pointed out when I wrote to him of this, parts of it were based on actual events. As always, I admired the writing—sinewy sentences, many of them bearing gifts of new words—but it was one of the last Naipaul novels I read. Happily, there were his travel books. As part of his becoming more English than the English, he had embraced—and would pretty much conclude—the long tradition of great British novelists—Lawrence, Huxley, Orwell, Greene, Waugh, Durrell—who were also remarkable travel writers.

I had time to read, but I wasn't as productive as I'd been in Greece. I still made drawings; a recent one was of a band of Warsaw street musicians based on a photo in my Polish textbook. I sent it to Hania, along with a letter suggesting I return and take up my old job at the English Language College. I added my belief that we should get married.

Unlike in Arta, here I had a social life. Bob Joffee was living nearby, in a carriage house on an estate, and one evening he invited me to join him at his French cooking class. Bob was a foodie before there were foodies. When the owners of the estate went on vacation he threw a dinner party out by the pool; after several bottles of wine, someone broke a crystal glass. Everyone except Bob found this very amusing. We half-heartedly threatened to break some more. It was vaguely Gatsbyesque but ersatz at the core: We were not wealthy; we were simply borrowing the scenery. We were *in* Princeton but not *of* it. Sub-simian. More on my level, I attended a costume party dressed as a blind beggar with a seeing-eye dog—Buffy, of course. I knew she'd love the drive—she sometimes sat in the car, enjoying the breeze, while Sally worked in the newsroom—and that she'd be perfectly at ease in a roomful of strangers. We drove back to Trenton, in that less sensitive time, carrying first prize.

Some Saturday evenings I would drive with Krystyna and her husband to Philadelphia. We would park in Old City and walk down Elfreth's Alley—now empty of tourists—to the half-open Dutch door of Kazimierz's row house. Kazio was an old friend of Krystyna's from Krakow, an architect who was also a gifted artist. The house was filled with his paintings, though I found the layout, with each floor consisting of a single room, as intriguing as the artwork. Gatherings always started in the living room and, eventually, ended in the half-subterranean kitchen, often with Kazio, his naturally rubicund face even more flush from vodka, belting out old communist ditties. "*Budujemy nowy dom, jeszcze jeden nowy dom . . .*" ("We are building a new house, still one more new house . . ."). It pleased me to be partying in a place my parents had dutifully shown me as a child—Elfreth's Alley is said to be the nation's oldest continuously inhabited street. The fact that I was doing so in Polish (sort of) only heightened my feeling of acquired cosmopolitanism.

I also welcomed visitors to Princeton. Jolanta arrived from Poland on her way back to the University of Michigan and spent a few days with me in the secluded house. She had not met Hania, but she knew all about her; she slept downstairs on the foldout couch. Friends of Hania's—Americans who'd been living in Warsaw—stopped by for iced tea on their drive back to Arizona. Carla had a thin, long-haired, hippie look—she may have been wearing granny glasses—and a sunny disposition; Robert was quiet and thoughtful with a beard and a pipe. They brought me cherished news of Hania, who had helped them as she had helped the Meinkes. With very few words of Polish and none of French,

Carla had gotten on famously with Marylka and even Elżbieta, the shy anesthesiologist whom I found aloof.

They filled me in on people Hania had been mentioning in her letters, like the English student Andy Mew and his wife, Mercedes, who worked at the Mexican embassy. Since Robert had been a student, they had lived in a dorm.

"The first week there was no light in the corridor," he said. "So I went out and bought a bulb and screwed it in myself." He seemed surprised that no one had done this before him.

A young woman from Madrid arrived at my brother Jim's house on a Rotary Club exchange, and, because I wasn't working, I was enlisted to show her around. It was my pleasure. Sonia had curly brown hair pulled into a tail that accentuated her El Greco profile. When she found herself at a loss in English, we switched to French. I took her to Great Adventure on an oppressively hot day, and as I sucked on the leftover ice cubes from my soda she told me that they would make me even thirstier. What was it with Europeans and ice cubes, I wondered? We had popularized them—why couldn't we do with them what we wished? I took her to the Hopewell quarry, with Dan and Jane, and tried not to stare at her bikini, her skin that seemed naturally tan. My tennis partner and his Portuguese wife invited us for dinner, and on the way I stopped to show Sonia the house where I was staying. Upstairs in the bedroom she sat on the bed, a move I thought might conceivably be an invitation. I was not astute at reading female signals, and when one seemed positive, as this one did, I tended—because of my many seasons on the sidelines of the romance game—not to believe it. I was horribly tempted to test it out, but I felt bound to Hania. Especially having just sent her a marriage proposal. The house—not the one we'd first slept together in, but owned by the same people—may have played a small part in my restraint. It was the age, past it really, of the sexual revolution, but I had been raised on the pop songs of the '50s and early '60s that had come before it—"Sexual intercourse began" wrote the poet, "In nineteen sixty-three"—and that had implanted in my young brain a possibly utopian ideal of true love. We headed out to buy a bottle of sangria.

The dinner was mostly pleasant, though at one point the grad student humiliated his wife, a sweet woman who seemed devoted to him. When he talked about his work, you could see that he approached the people he studied with true sensitivity; and, when discussing other subjects, he seemed curious and reasonable, if a bit naive. He didn't understand why the Poles didn't just say they'd had enough of the Soviets and go off on their own. Despite this almost childlike approach to reality, at least the geopolitical one, he was extremely smart. So I was not only bothered but also confused by his treatment of his wife. I had seen him, on other occasions, fly into rages directed solely at her. It astounded me that some-one possessing such intelligence could be so lacking in self-awareness. I eventually discovered that there is often no connection between the two.

When Sonia heard that I was going to New York—I had applied yet again for a Polish visa—she asked if she could come along. I'm sure Manhattan was more the draw than I was, especially for a Madrileña who had been stuck in Warren County, but, when I came to pick her up and she burst from the house fresh from her morning shower in a hoop skirt of boldly colorful stripes, my mind flashed back to the untried bed in Princeton. Yet how would I have felt, had we tried it, heading with her now to the consulate of my love?

Once again I was given a visa—and this time I wondered, irrationally, if it was my reward for being faithful.

My listless summer was coming to an end. A letter arrived from Random House; Ms. Glowczewska graciously wrote that she understood how the experience of Poland must have seemed exotic to a young Amer-ican like me, but unfortunately she didn't think the novel was going to work. Since she so obviously saw it as autobiography, I had to agree with her. It was the last time I would attempt to write fiction, but it was not the last time I would be rejected by Ms. Glowczewska who, seven years later, would join the staff of the new *Condé Nast Traveler* magazine.

The galleys of the book also arrived and, reading them before sleep-ing, I panicked: The rhythms of my sentences all seemed to be off. It was the first time I had proofread words of mine that wouldn't be fleeting, and the idea traumatized me so much that, as it turned out, it rendered me incapable of reading them properly. But that night I sat on the bed in a nervous sweat.

One day, stopping by Mike and Beth's house to feed their cat, I turned on the TV and found Dick Cavett interviewing Jan Morris. She was talking about her favorite travel writer, Alexander Kinglake, and his nineteenth-century classic, *Eothen*. Both names were new to me. That I was learning about them from a television talk show seemed not nearly as strange then as it does today. I added another book to my unending list.

A week later I read in the newspaper that shipyard workers in Poland had gone on strike.

By the end of the month, the *Washington Post* was rich in stories with a Gdansk dateline. I was spending a long weekend in suburban Virginia with Blaine, who seemed unmoved by the fact, pointed out by one of the analysts—Neal Ascherson, perhaps—that uprisings in Eastern Europe occurred in twelve-year cycles: 1956 (Hungary), 1968 (Czechoslovakia), 1980 (Poland). He was much more impressed that I was going to be in the middle of the latest one.

Shortly after he headed off to the newsroom, I went to a nearby diner for lunch. My waitress whined to a colleague that she had no plans for the weekend other than hanging out with her boyfriend. She made the coming few days sound like a sentence. How many weekends with Hania had I missed out on? I wanted to grab my waitress by the shoulders and, Lucinda-Matlock-like, tell her to cherish every moment with her boyfriend.

Back home in Phillipsburg, I packed the big leather suitcase, this time even heavier because of the copy of *New Jersey: Unexpected Pleasures*. The subtitle, considering the uncertainty into which I was headed, seemed a little bland. But the book was attractive. My "Bridges" essay appeared first and carried a picture of the covered bridge in Hunterdon, here softened by snow, where I had reflected before my first trip to Poland.

Familiar with the drill, my parents drove me to Kennedy in my father's black Buick. From the backseat I told them I felt more confident going to Europe this time than I had on the three previous occasions. Their silence told me they were filled with even more concern.

CHAPTER 5

Warsaw Found

"You've come at a terrible time."

This was not the greeting I had been expecting.

Nor the assessment. Reading at home of the birth of Solidarity and watching the news—seeing the intense, mustached Lech Wałęsa addressing intense, mustached workers—had filled me with excitement (something was happening) and hope (predictably). But Hania's dour words suggested that revolutions are often less attractive the closer you are to them. For all my disappointment at her tone, I tried to take comfort from the idea that, by being here, I was going to see things differently than people in the West.

Hania's face was drained of color and taut with worry; her hair, which she had let grow, appeared stringy and unwashed. It had been a year since I'd seen her—the longest we had been apart—and I wondered if the image I had kept of her had been deficient, airbrushed with longing.

We took a taxi to her new apartment. It was not far from the airport, in a small development in the unprepossessing district of Ochota. Marylka greeted me warmly; Elżbieta was less enthusiastic, due to her shyness or perhaps her realization that there would now be even more competition for the bathroom. We piled into the kitchen, where Jaś, Elżbieta's boyfriend, rose from his low, square stool, and a German shepherd manqué, a replacement for the recently departed Babichon, walked over to sniff me. Seated anew, Jaś gave me a tutorial in Polish on different types of boats. I knew this because his words were accompanied by drawings. It was odd to think that a revolution was brewing. We all drank tea, and I was served cold fish in tomato sauce, which I told Marylka was "*niebo w gębie*" (heaven in the mouth). It wasn't; I simply wanted to be nice and to impress them with an expression I had learned over the summer.

Hania and I eventually retired to the main bedroom, where a bed without a headboard was set against a wall and a Biedermeier desk stood expectantly by a balcony door. I opened my suitcase and unearthed my copy of *New Jersey: Unexpected Pleasures*.

"It's beautiful," Hania said, paging through the photographs, and for the first time I felt pride in the book.

We took turns using the bathroom, then everyone retreated behind closed doors; there was one other bedroom, and Marylka slept on the divan in the living room. Each small room became a personal refuge. Hania and I climbed into bed and, very quietly, began to make love. It had been a long time. She started to cry.

I hoped it was out of happiness, but that wasn't like her.

"I'm not in love with you anymore," she confessed sadly, as more tears flowed.

After the shock I made a dumb joke, which caused her to laugh. It was one way to go. But I was devastated and caught completely by surprise. I could understand someone falling out of love with me when I was around, all my faults and shortcomings in daily view. But how do you fall out of love with someone who is thousands of miles away? If you don't find someone else, as Hania assured me she hadn't. That had been my concern; not that my absence would result in erasure. I had always thought my absence would work to my advantage, that the idea of me— aided by my written words—would be more appealing than the reality.

With my heart wounded, my mind reeled. My confidence in returning to Poland now looked idiotic; the plan, it was clear, had been a colossal mistake. My future, as I had envisioned it, no longer existed.

Yet Hania insisted she wanted to marry me. As the one still in love, it was not for me to say no.

In the morning we joined everyone at the kitchen table, my dumbstruck sorrow masquerading as jet lag. A cheesecloth, I now noticed, hung from the window, which gave onto the gray wall of a similarly grim block set perpendicular to our own. At the Frankfurt airport, I had run my eyes lovingly over the bright wrappers of candies, the colorful bags of chips, knowing I would not see gorgeously packaged junk food for many months. But that was, I had thought then, a small price to pay for love.

Fall had already arrived, cold and damp, and was so thorough in banishing summer that it was difficult to remember the dusty warmth of my arrival two years earlier. Which is why, when I wrote my book about Poland, I stated in the opening sentence that I had first arrived in the fall. Fall, this fall, with its low-hanging skies melodramatically presaging the darkness to come, seemed to me, especially now, to be the defining season of Poland—its abiding mood autumnal.

The weather made my new neighborhood look even sadder. To get to the tram I had to walk several long blocks along Opaczewska, a street that had inspired a patriotic poem because it led to the spot on Grójecka Street where a group of soldiers and valiant citizens had tried to stop the advancing German army in the fall of 1939, a fall—as Marylka often reminded us, with the hapless despair of the perpetually forlorn—that was improbably balmy and clear.

Despite its place in verse, the street barely qualified as prosaic. There was a corner bakery, a dark cavity with a worn wooden counter where residents bought loaves that looked like footballs that had been rounded at both ends. Our daily bread. There was the perpetual gloom of the *spożywczy* on the other side of the narrow, overgrown park that bordered the street, where sullen women in dirty aprons and bulky sweaters shoveled our potatoes off a soil-darkened floor. (Though this scene helped me remember the word for spud—*ziemniak*—because it has its root in the word for earth: *ziemia*.) Closer to the intersection, there was a string of small shops fronting minimalist window displays. There was no neighborhood spirit—just preternaturally focused shoppers—and there was no complaining, not when you stood in the rain waiting for a tram in the place where brave citizens died in defense of their country.

Now four decades later, there was a new crisis, a raw sense of urgency. The political situation took my mind off—or at least kept me from obsessing about—my personal life. Solidarity's free trade union, which in the States I had thought was relegated to shipyards and factories, had infiltrated most of the country's institutions; we soon had a Solidarity chapter at school, with Krzysztof as its leader. I wanted to join, but Hania vetoed the idea; as a foreigner, she said, I would be expelled from the country if the movement were crushed. She was already envisioning a bad ending, which was, despite my fresh romantic bust, almost unimaginable to me. I naturally assumed things would get progressively better.

I loved going to school, talking to the newly energized teachers, teaching the eager and receptive students. I would enter the scripture-engraved building on Plac Zbawiciela—fittingly, it was the line from Matthew: "Come unto me all ye who travail and are heavy laden, and I will refresh you"—and climb the circular staircase to the teachers' room, which was louder than ever with news and discussion. Then, at the bell, I would head off to my classroom. I had one large, interesting intermediate class, with a lot of characters—current events only bolstering their strong personalities—and another large and delightful class of beginners, which included a married couple, Anna and Tadeusz, who insisted on being called "Ted." Their smiling, earnest faces would have put even a novice teacher at ease, but the fact that I now had experience—in this school, in this classroom—gave me not only confidence but also the capacity for enjoyment. In life, with the exception of repeating a grade, which nobody does voluntarily, we almost never get to do something important all over again, and I had been given that rare opportunity. Along with the front-row seat at the revolution, this gift helped make more palatable my decision to return.

Also, in the realm of good timing, my return coincided with the meeting in Warsaw of the Society of American Travel Writers. I was not a member; I had not published any travel writing; my "Iron Curtain Local" story had been roundly rejected. But I had been corresponding with Caskie Stinnett, who, as a past president of the organization and a former editor of *Holiday*, was the doyen of American travel writers (periodicals division). *Holiday*—along with the *Saturday Evening Post* and *Ladies' Home Journal*—had been put out by the Curtis Publishing Company in Philadelphia and been hailed for both its design and its content, which was provided by some of the best novelists, short-story writers, and humorists of the day: William Faulkner, John Steinbeck, James Thurber, Eudora Welty, Evelyn Waugh, Paul Bowles, V. S. Pritchett, S. J. Perelman.

A request from the editors to Patrick Leigh Fermor in 1962, asking him to write about "The Pleasures of Walking," was the spark that finally got him to begin his account of his trek across Europe. E. B. White's long essay "Here Is New York" was written not for his longtime employer, the *New Yorker*, but for *Holiday*. Our Jersey-shore-going family had not subscribed to the magazine—I got my knowledge of the world from *Life*—and I discovered it, with a fortune hunter's delight, in a dusty box in a secondhand bookstore.

Stinnett had ended his editing career three years earlier, at the much less ambitious *Travel + Leisure*. Travel magazines were in decline but writers like Theroux—who had followed *The Great Railway Bazaar* with *The Old Patagonian Express*—and Bruce Chatwin (*In Patagonia*) were giving new life to the travel book. I had written to Stinnett that I would be living in Warsaw during the SATW meeting, and he had kindly arranged for me to attend.

It was the strangest week I had ever spent. In the morning I would stand in queues for food—the shortages were seen as a government ploy to get people to turn against Solidarity—and in the evening I would sit down to feasts in luxury hotels. One evening at the Victoria, a fashion show accompanied our dinner, the models grinning down at us from their runway heights. More striking to me than their beauty—they could have been my students, with lots of makeup—were their expressions. Never before had the divide between the world of the tourist and the world of the resident appeared so stark.

There was an odd dynamic to the meals themselves, as I found myself in conversation with people who were engaged in the work I wanted to do yet were, for the moment at least, envious of me. In the destination of the travel writer, the expat is king. They were curious about what was happening in Poland—suspecting that three-course meals graced by sashaying houris weren't exactly the norm—and peppered me with questions. The change in status from benighted foreigner (except in the classroom) to fluent expert was so enjoyable that I invited home one young freelancer from Alaska who came along with her husband, a man who had quit his job to become a prospector. Weeks later Hania, trying unsuccessfully to remember the word, described the man as a "golden retriever."

The closing dinner was held at the Forum in the commercial heart of Warsaw. Perhaps because it was rumbled by trams on the city's two major boulevards—Jerozolimskie and Marszałkowska—the hotel always had for me the slight air of a way station. Though you would not have felt it entering the banquet hall this night. Gazing at the food sculptures crowning the tables, I thought of the potato-piled floor of my local *spożywczy*. I actually felt guilty, partaking of such excess while the world beyond these walls struggled with shortages. The fact that I was now part of that world, with only this brief reprieve, made little dent in my discomfort. The whole glittering gala seemed an indictment of travel writing, at least the kind these people practiced. It was not just that they were ignoring the realities of the place they were visiting, shutting themselves off in a sumptuous dining room, but that, by indulging in forbidden luxuries, they appeared to be thumbing their noses at them. For all their expressed interest in Solidarity, these good-time travel writers were attending an event that represented its antithesis. Some of them may have also recognized the inappropriateness of the event and felt a similar sense of guilt or at least uneasiness. But that dinner gave me my first glimpse of why most journalists regard travel writers with disdain.

My meeting with Caskie Stinnett didn't come until the end of the evening, when he was standing at the coat-check counter. I introduced myself and, after some small talk, asked him what he thought of the current *Travel + Leisure*.

"I've learned," he said, putting on his overcoat, "that it's best never to look back at a magazine you once edited."

I emerged from the week more determined than ever to learn about Poland and enrolled in Polish classes at the University of Warsaw. The textbook was written for use by students from three different language groups, a strategy that pretty much guaranteed that we would all, in coming to the glossary at the end of each chapter, question our decision to try to learn Polish. *Przyjaciel* was followed by *friend, ami, amigo*; *księżyc* by *moon, lune, luna*; *skrzypce* by *violin, violon, violín*.

The only other students in my class were an American couple, Mary and Paul, old socialists from Colorado who were spending the year on

Paul's academic scholarship. When not trying to express how terrible everything was in the United States and how excellent everything was in Poland (from bread to television), they struggled with the various verbs for "to go"—on foot? by vehicle? how often?—especially when entering the vast maze of prefixes. One day, frustrated at their continued inability to correctly say where they were going that evening, I shouted out, "*Lepiej zostać w domu*" (Better to stay home).

I had changed my attitude toward Polish; instead of fighting it, as I had on my first go-round, I decided to embrace it, which is the only way to learn another language. Now, rather than moan about the grammar, I tried to savor the fact that you walked down the sidewalk to the *biblioteki* and then read quietly in the *bibliotece*.

I eventually stopped going to Polish classes, convinced that I could learn the language faster on my own. Sitting in the kitchen and listening to Marylka, who had wanted to become an actress, was educational, and Jaś turned out to be a patient teacher. An engineer by profession, he had a deep knowledge of Polish history and an equally profound desire to share it. Hania, still not convinced that I needed to learn Polish, was of little help. Because we had met in London, when her English was already fluent, English was—and hopefully long would be—our language. And we obviously didn't need the extra tension in our relationship that results when one partner tries to teach the other a new skill.

Lists of Polish words grew on my yellow legal pad, which joined the one for English vocabulary. My life had become a life of words: teaching them, writing them in sentences, writing them in columns (before memorizing them), reading them in books, reading them on screens. Unlike everyone else in the apartment, I loved when Russian movies were shown on TV, because they came not only with a speaker—a man whose voice remained unchanged regardless of the gender or emotions of the actors—but also with subtitles. So I could read on the screen the words that were being spoken by the speaker. "Take that," he would say matter-of-factly during a brawl; love scenes were even more implausible. But he was the perfect Polish-as-a-second-language instructor. I'd sit with pen and paper, writing down unfamiliar words, and then, after the movie, I'd look them up in the big Polish-English dictionary—*Wielki*

Słownik Polsko-Angielski—that Hania had received, according to the inscription on the title page, for excellence in her classes at Szymanów.

There was a Marek Grechuta album in the apartment that I listened to often, though, regrettably, it came with no accompanying sheet of lyrics. I had made a clean break from pop music in 1974, the year I graduated from college and moved to Washington, a city with an excellent classical music station. It was time, I had decided, to elevate my musical tastes. In France, I had started listening to contemporary music again, but it was different from American music of that era, still very much in the singer/songwriter tradition. The soundtrack to almost every party in Aix had been the music of Georges Moustaki, whose words and melodies seemed designed to make you want to drift off into bed—"to jump" would have been far too strenuous for anything inspired by his somnolent purr—with whomever you were talking to. My teacher had taught us the lyrics to a few Georges Brassens songs: "*Le Parapluie*" ("The Umbrella") and "*La Première Fille*" ("The First Girl")—the national troubadours perfectly perpetuated the global image of the sex-obsessed Frenchman. My girlfriend Martine had introduced me to Léo Ferré, who sang the poetry of Aragon and Baudelaire, and Jean Ferrat, whose "*C'est Beau La Vie*" seemed a robust Gallic answer to Louis Armstrong's "What a Wonderful World."

Grechuta was different; he was Polish, for one thing, and from Krakow, the home of another great contemporary singer, Ewa Demarczyk. The song of his that first caught my ear was "*Wesele*" ("The Wedding"), which, Hania explained, put to music words from a play by the nineteenth-century poet, playwright, and painter Stanisław Wyspiański. So, like Ferré, he made use of literature. But his arrangements were more complex, incorporating violins (*skrzypce*) and other instruments not normally heard in popular music. They made the songs of James Taylor, Gordon Lightfoot, and Leonard Cohen, which I loved, seem rather simple. I listened over and over again to the way he said the word for bride—*panno młoda*—distinctly pronouncing both of the n's—and then turned it around—"*młoda panno*" (young lady)—making what I assumed was a pun.

Browsing in a secondhand bookstore one day, I came across a book with the name Zanussi printed in black letters up the spine. There is no rule in Polish publishing regarding the positioning of words on spines, which means that, in a bookstore, customers continually have to move their heads from side to side to read the titles. I found this more than irritating; it was a sign not of a charming disdain for uniformity but of a national lack of exactitude and a deleterious inability to come to an agreement—which was reflected in the country's contentious political history.

Opening the book and going to the table of contents in the back—another infuriating Polish publishing practice—I saw that it contained eight screenplays of the director's films, including one, *Bilans Kwartalny*, that I had seen in New York under the title *A Woman's Decision*. I immediately took it up to the register. In France, I had helped my French along by reading a book of screenplays of four Truffaut films, sometimes committing short dialogues to memory. If you are trying to learn to speak a language quickly, novels are useless because of all the description and expository writing. Even plays are of minimal help, because the language is often stilted and didactic. A screenplay, at least one of a contemporary movie, contains the everyday speech of everyday people.

And I wanted to learn Polish as quickly as possible, because Poles were having intense discussions, reading the newspapers and debating what they read. I desired, as any sentient person would have, to know what was in the air around me. Also, as with teaching, I had been given a do-over with Poland, and I was determined not to squander it.

My progress in learning Polish was momentarily halted by my discovery, in a bookstore in Żoliborz, of Christopher Sykes's biography of Evelyn Waugh. It sat like a paperback mirage on a table with mostly Polish books. One sometimes came across translations of novels by John Steinbeck and Jack London, but a biography of Waugh—a first-class snob and class-conscious curmudgeon—looked completely out of place. I carried it like contraband to the counter, though later it occurred to me that the great writer, like many members of Solidarity, was a practicing Catholic and a fervent anticommunist.

I also returned to the British Council on Aleje Jerozolimskie, now as much to read another take on current events as to lose myself in novels and travel books. There was a magazine, the *Spectator*, that had a correspondent in Poland who was as dazzling a stylist as he was an analyst. Timothy Garton Ash knew his history and recognized the importance of the Catholic Church (which already distinguished him from many of his ardently secular colleagues), and he wrote with not just intelligence but flair; in his articles, the old clichés about Eastern Europe were replaced by inspired and frequently irreverent insights. I could enter the Council feeling depressed about my life in Poland and then, after reading his latest dispatch, think it the only place to be.

He wasn't the magazine's only spirit-lifting voice. Its pages were packed with writers who had a contrarian wit and a jaunty self-confidence built on deep, if lightly flaunted, stores of knowledge. Someone once described the magazine as "cheerfully Tory anarchist." It was a home for gifted columnists, including Waugh's son Auberon, who seemed to delight in taking positions at odds with their colleagues' a few pages away. There was a "High Life" column, carrying the byline of Taki, whose self-assurance was built on his Greek family's shipping fortune, and next to it, surprisingly—if appropriately—a "Low Life" column, written by a man named Jeffrey Bernard, who seemed to divide his time between the track and his pub and, weekly, turned his marginal life into minimalist art. Whenever he failed to produce a column on deadline, the note appeared: "Jeffrey Bernard is unwell." Nobody like him—nothing like the *Spectator*—existed in American journalism, and the magazine sharpened my taste for the English, at least as they appeared in print.

But I still frequented the American embassy library, and on teaching days I would stay for the lunch that was served in the basement. Cooked by the Polish staff, it generally consisted of watery soup, stringy meat, boiled potatoes, and sometimes a side of grated cabbage or beets. It was the one embassy privilege that I—a private citizen with no "official" purpose in Poland—was granted, and it was a cherished one; the opportunity to eat food that one hadn't had to search for and then spend a morning in line to purchase had become a great gift.

The cafeteria was also a good place to meet people, not so much embassy staff—as they tended to keep to themselves—but the motley of Americans who, for various reasons, had landed in Poland. There was Bill, a young ethnomusicologist from a Midwestern university who would disappear for weeks, visiting villages the rest of us would never see. And there was Roman, a graduate student in political science who had come to write about Solidarity. Serious and intense, he would rarely linger, heading off as soon as he had finished his lunch to what we assumed was an important assignation.

I wasn't entitled to shop in the embassy commissary, though one evening I got invited to the embassy bar, located next door in the residence that housed the Marines. We ordered beers from a striking young Polish barmaid—another!—and got talking to a Japanese college student who had come to study the works of Witold Gombrowicz. An older American sitting at the end of the bar insisted that the key to Polish was its alphabet: Once you learned how the letters were pronounced, the language was a cinch. This seemed at the time, and eventually proved to be, a vast oversimplification of a basic truth. I feasted on fistfuls of salted peanuts.

A few hours later, as I boarded the #15 tram on Marszałkowska Street, a pretty young woman smiled at me. When I looked perplexed—smiles on public transport were as rare as grapefruits—she explained that she was the barmaid who had just served me beers. For a moment, Warsaw seemed like any great city, an intricate web, a coincidence realm, until I remembered that many of the Poles who worked in Western embassies were thought to be informers. The bartender's smile was like a finely wrought "welcome back."

On a drizzly Saturday morning in October, I woke to find Hania hovering over our bed. "I'm sorry," she said softly, "but you have to get up and get married."

We both had colds and dressed amidst sniffles; me in one of my tailor-made, black-market suits and Hania in an off-white dress with a loose belt sporting a flower at the hip. She wore a gold necklace and carried a hand bouquet of fifteen yellow roses. Her hair was tied in a bun that showed off her neck and profile beautifully. I should have felt like the luckiest man in the world.

A relative I had never seen before piled us into his car and drove us to the Old Town and the Palace of Weddings. A few of Hania's aunts and friends were already there; my only support came from Krzysztof, who was serving as best man and interpreter. After a brief wait, we entered a room with two small windows and took our seats in front of a large table. An expressionless female magistrate soon appeared, in a cherry-red robe and a large chain necklace that dropped a crownless Polish eagle onto her bosom. We repeated our lines—Hania in Polish, I in English—and exchanged rings; the one I'd worn to Greece found my finger again. As Mendelssohn's "Wedding March" played, we exited to form a modest reception line, where we were each kissed three times by men and women. Back at the apartment, Marylka greeted us at the door with a loaf of bread topped by a saltshaker, an old Slavic custom that supposedly ensures that the newlyweds will never lack for the essentials—at least the material ones. Hania hung her bouquet from the ceiling lamp in our bedroom.

It would be hard to imagine a less promising start to a future together: living in a loveless marriage—or, rather, one of unrequited love—and sharing a cramped apartment in a broken-down country with family members who were not real family.

In the evening, a small group of us went for a quiet dinner at the Cristal Budapest. Jaś gave the roving musicians money to move to another table. Krzysztof talked to me about the Nobel committee's decision, announced a few days earlier, to award the prize for literature to Czesław Miłosz. Because the poet was not well known, he thought the award was politically motivated. I expected Krzysztof, like most Poles, especially those supporting Solidarity, to be thrilled by the news, and I put this criticism down to his skeptical side, his refreshing need to question popular opinion.

Monday, walking into my classes, I was greeted with smiles and whispers and a-ha looks; the female students, especially, had noticed the ring.

A few days later a large envelope arrived from the States containing the only wedding present I would receive: *Winter in Moscow* by Malcolm Muggeridge. It had been sent by my friend Phil Terzian, who had been a year ahead of me at Villanova. Phil was a formidable figure on campus, as commanding, and seemingly as well read, as many of the professors. We had both worked on the college newspaper, though we had barely acknowledged each other. But in Washington, where Phil was working at the *New Republic*, we became friends. He was still high-handed in his opinions; once, riding in his car down Connecticut Avenue, I had expressed admiration for Christopher Isherwood—*Berlin Stories* had made me ravenous for foreign experience—and been told he was a laughably bad writer. But Phil always treated me kindly, perhaps relishing our mentor-pupil relationship. He was a devoted Anglophile and introduced me to Max Beerbohm, whose essay collections I plucked from the shelves of secondhand bookstores in Georgetown and Alexandria and read with a delight that, I suspected, would have been lost to me in graduate school—as Max would have been. Scholars must study the great, while writers often learn more from the merely sublime because their lessons are more applicable and their heights, seemingly, more attainable. On Phil's recommendation, I had read both volumes of Muggeridge's memoir, *Chronicles of Wasted Time*, and been taken as much by his sensibilities—world-weary, amused by life's absurdities, tentatively devout—as by his dismissive, if perfectly English, labeling of a distinguished and checkered life. I would be thrilled, I thought, to similarly squander my days.

Phil had left the *New Republic* and become the associate editor of the *Lexington Herald* in Kentucky. He was a fan of the South, not just as a child of Maryland but also as a conservative and the husband of a woman from Nashville. (The year before I had been a groomsman at their wedding, which had been attended by Minnie Pearl.) In his note, he urged me to send him occasional op-ed pieces. There were now quite a few foreign correspondents in Warsaw, but none of them were married to Poles (yet) and none were working and living with them. As someone who had never wanted to be a reporter, I found myself in an ideal situation, at least professionally. My front-row seat was not at the main event, which was in full view—or as full as it could be in a socialist country—and was being written about extensively in numerous languages, but off to the side, on the margins where history is not made but lived. I was more Isherwood than Muggeridge, a camera observing a people in extremis. Yet Muggeridge, a correspondent for part of his career, recognized the folly of making politics paramount. "He came to believe," his biographer Ian Hunter wrote, in a passage I used as the epigraph for *Unquiet Days*, "that the machinations of power, how it was organized and wielded, who governed whom and by what means, all this was less important than a nation's soul; its character; its religion; its humour and art and music and literature."

This, I had decided, would be my bailiwick.

Though it didn't preclude my hanging out with foreigners. One night after classes I went to the Hotel Europejski to meet the correspondent for the *Washington Post*. Owing to my days in Trenton, we had a number of friends in common. At the reception desk I asked for Bradley Graham—a more perfect name for a *Post* reporter could not have been invented—and the receptionist turned and pointed to the young man typing rapidly in the curtained cubicle behind her. It was not how I had expected to find him, tucked away in a cramped space behind the hotel reception desk, and it seemed to drain the words "foreign correspondent" of much of their glamour. Though for real journalists the sight might have added to it, demonstrating the adaptability and bare-bones dedication of the true professional. (He was far removed from the food sculptures of the Forum.) Not a friend of typewriters, I would have found writing on one in a public space nearly impossible. For the first time since my arrival, I thought fondly of our apartment in Ochota.

Bradley finished typing a sentence and came over to meet me. We talked a bit about mutual acquaintances and our current situations, and he told me about a party he was having with some friends in a few weeks. There must have been something so forlorn about his cubicle there behind the reception desk that I asked him:

"You have friends here?"

Also, in my defense, it was still early in his posting. But it was a stupid thing to say, and I realized this as soon as the words were out of my mouth. It came across as not just critical but boastful, as if I were the one who had established a home in Poland. Underlying it was my own feeling of inadequacy at being there as a teacher—the last refuge of an expat—and not as a journalist, for all my self-assurances to the contrary. We had both come out of Trenton and now found ourselves in Warsaw—but leading very different lives.

I had less in common with Hania's English friend from the university, but Andy and I quickly became friends. He was tall and handsome, with angular features and straight black hair; walking down the street with him I'd watch, enviously, as he'd catch women's eyes. We both stood out as foreigners, but for me it was mainly by dint of my round tortoise-shell glasses.

Andy was soft-spoken, easily amused and rarely ruffled; he took life with a confident ease that reminded me of the *Spectator* writers. Of course it helped, in Warsaw, that he was married to a woman who worked in the foreign service of a Western country.

Through Andy and Mercedes we met people from other embassies—Dutch, Swiss, French—none of them in high positions. But dinners and parties in their well-stocked apartments were pleasant, temporary returns to the world of comfort and plenty I had left behind.

Our downstairs neighbors, the Michałeks, would sometimes have us over. While Polish, they had a laidback informality that always made me feel good because it seemed so familiar. We would gather in the kitchen—a room that, except during parties, was usually off-limits to guests in Poland—while Kasia threw together an invariably delicious meal and Hania and Kubuś caught up on gossip. They had studied together at the University of Warsaw, where Kubuś now taught economics. His father was a well-known film critic, a friend of Wajda, and the family had spent a fair amount of time in the West. Dinner was always accompanied by wine.

As Christmas approached, the food situation worsened. A cartoon on the cover of *Kultura* showed a turkey lamenting, "My great-grandfather was stuffed with truffles, my grandfather with chestnuts, my father with tomatoes—I will have to be content with eggs." I felt bad that I wasn't doing my part in the grocery department, and expressed my concerns to

Hania. "It's not food shopping now," she replied, "it's food hunting—and I don't think you have the talents."

One evening we took a bus to the Old Town to watch a new documentary about Solidarity. I loved going to the movies in Poland because they provided an escape not just from place but from time. Black-and-white newsreels preceded every film, and this evening we watched as footage from Stockholm showed Val Logsdon Fitch, the husband of my old *Trenton Times* colleague Daisy, receiving the Nobel Prize in Physics.

During the documentary, Hania kept checking her watch. As we walked to catch a bus, I asked if spending time with me was so unpleasant she needed to count the minutes. She took offense, explaining that it was a habit of hers. We rode home in silence.

The season lifted my mood a bit. With the exception of carp, I loved everything about Christmas in Poland: the holiday party at school, when the table in the teachers' room was set with a decorative cloth and cuttings of pine for a high-spirited tea; the carol singing with students in the downstairs lobby; the sitting down to dinner at the appearance of the first star; the extra place setting on the table for the passing stranger—who, according to the joke this year, would be a Russian soldier; the faintly sacerdotal breaking of bread and the seemingly endless bestowing of good wishes; the economically mandated simplicity of gift giving; and, of course, the beautiful Polish carols that were made all the more meaningful by never being heard in stores.

On New Year's Eve we celebrated at a "pink party" given by Ineke, one of our new Dutch friends. Hania put on a pink dress and, using her nail polish, painted my tortoiseshell glasses pink; then she found a pink belt of fabric that could double as a tie. We took a taxi to Ineke's apartment and, getting off the elevator, entered a room full of Europeans in pink. The only other American in attendance was a Marine, who spent the evening in a corner making out with a young woman from Finland. Actually, there was an American from the southern hemisphere, the voluptuous daughter of the Paraguayan ambassador, who described her country's embassy as "my father and this other guy."

One middle-aged man wore a pink sash rather dashingly across his suitcoat; it was, he explained, part of his former shower curtain. Later in the evening, he was seen dancing exuberantly in a Polish mountaineer's hat. Hania at one point ended up next to him and asked him casually where he worked.

"At the Dutch embassy," he said.

"What do you do there?" she asked.

"I'm the ambassador."

At midnight we all raised our glasses. There was a certain luxury to ringing in the new year with people who could go home if it turned nasty.

Hania and I were driven back to Ochota by the factotum at the French embassy. Albert's son was dating a woman who worked at the Dutch embassy, though he had spent most of the party sulking alone in a haze of Gauloises.

In his low-riding Citroën, Albert wondered if we had noticed his son.

"I ask you," he said in French, "What am I to do? He is completely blasé. He has no desire, no idea what he wants. He is my son and yet I fail to understand him. He lives with Anneke but he doesn't think of her. He doesn't talk to her. I can't understand it. For me, a woman is something sacred."

We had no advice for Albert, except of the directional sort, as he repeatedly drove his Citroën onto the sidewalk. When we eventually reached home, Hania expressed surprise at his last words; French men, she had long thought, were impenitent philanderers. I reminded her— she whose native tongue is devoid of articles—that he had said, "*une femme est quelque chose de sacré,*" not "*ma femme est quelque chose de sacré.*"

1981 began with the now customary strikes and threats and rumors, worsening shortages and reports of police actions. At school, we began receiving monthly ration cards for meat, rice, sugar, and butter. A rare letter arrived from my father, urging me to return home with Hania; he could not stop thinking, he wrote, of September 1, 1939.

But the weather, while dreary, was not apocalyptic. It was as if, because everything else had become so dramatic, it didn't have to be. My long walks home along Opaczewska were still dispiriting—I cursed the cold and the closed-up shops, a city deadened by the system as much as by the season—but I never had to worry about muggers.

At the end of February, we packed Jaś's Syrena for a trip to the Tatra mountains. The piling on of provisions—canned goods, teas, medicines obtained by Elżbieta at her hospital—gave the journey the feel of an expedition. The thinking was that however bad things were in the capital, the situation in the provinces was even worse. Thick blankets added to the pioneer aspect.

South of Krakow, the road ascended into a wintry realm. Muddy streets were replaced by snow-covered forests. The gray concrete blocks of the city gave way to angular wooden houses. I had never spent significant time in the mountains, and I was struck by the feeling of separateness that was even more pronounced, I suspected, in a country where the soiled and dilapidated had become the norm. Polish mountaineers (*górale*) were very much their own people, I'd been told, and felt a bond to the Tatras that nearly superseded their ties to their country—a bit, in that sense, like Texans and Texas. Indeed, when Poles talked about the mountaineers, usually with admiration for their artistic nature or musical talents—or even their keen business sense—they sometimes used the words *zdolny naród* (gifted nation). When the pope had visited, returning to the mountains in which he had loved to hike, he had been serenaded with a well-known song containing a plaintive, pleading refrain:

> *Góralu, czy ci nie żal?*
> *Góralu, wracaj do hal!*

(Mountaineer, aren't you homesick?
   Mountaineer, return to your mountain pasture!)

We stayed in a guest house and took long walks over snow, Hania and Elżbieta following Jaś and me. I didn't like the arrangement—I wanted to be a couple, on our own, as we'd been in Trenton—but the scenery proved a bright distraction: sturdy houses of wood and straw; fluffy white dogs, similar to Great Pyrenees, that stood sometimes in front of their own mini-chalets; the carved wooden figures that combined piety and whimsy, even the popular and nationally apt "Christ Worrying," which showed the savior seated on a stump with his head tilted woefully into one hand; the soaring, discordant fiddle music, which we heard one day at a wedding before watching the groom and his bride—*panna młoda*—get carried off in a horse-drawn sleigh. We had entered another dimension, a place suspended in time. On Sunday morning we went to a church in the village of Chochołów and heard the congregation—men seated on the left, women on the right—sing a cappella before the mass, their rough, untrained voices creating a beautifully piercing harmony. On our last night in Zakopane we watched as men in traditional dress—tapered white wool trousers and black domed hats—emerged from the darkness to attend a service for the state of the nation. It was a sober if colorful reminder that we had not left Poland after all.

As soon as we returned to Warsaw and I'd gone through letters from family and friends, I started writing about the trip. Even though I was living abroad, I had created a kind of home in Warsaw, and leaving the capital, seeing another part of the country, witnessing ancient pastoral rites, had inspired the travel writer in me. Also, I had the feeling—already rare in travel writing—that I would be describing something that was unfamiliar to my readers. If I had any.

On days when I didn't teach, I wrote in longhand at the Biedermeier desk in our bedroom. In the lede I attempted to point out the uniqueness of the Tatras, comparing them favorably to ski resorts in the West, none of which I had ever seen. This relatively straightforward beginning was spiced, or marred, by an arch tone I had picked up from some of the Brits I'd been reading. I chided an anonymous American travel writer—perhaps one of those who had convened in Warsaw several months earlier—for complaining in a ski magazine that the region's chairlifts "were terribly inferior." But what, I asked, "of its folk dances compared to those of Aspen? Or its wedding rituals seen next to the divorces of Vail?" It was my attempt to turn the Polish joke—which I was now seeing as not only unfunny but unjust—back on the country that had created it.

The rest of the piece was less combative, focusing on the mountain culture. I included the names of famous families of the region and, taking my cue from Leigh Fermor, elaborate descriptions of the distinctive dress.

And, because I wasn't writing on assignment, to an arbitrary word length, I wrote as much as I pleased, working in seemingly tangential things, like a passage from a book of communist propaganda I'd been reading and a joke about Lenin. I was taking full advantage of travel writing's all-inclusive nature. I changed the time sequence—three years before Alistair Reid would admit to doing the same in the "Letters from Spain" he wrote for the *New Yorker*—ending the story with the wedding instead of the service for the state of the nation, as the former worked better with the quote from Hilaire Belloc I wanted to use. It was a travel essay more than a travel story; there were no encounters with locals, merely observations of them, and there were no dialogues, not even among the four of us. But for some reason I didn't see this as a failing. It was a passive account told in an impassioned voice; in a country of tradition under threat from a foreign power, the most traditional citizens—the mountaineers—were proudly and defiantly holding on to their old ways. The Tatras reinforced the idea I had already gleaned from living in Warsaw: that tradition is not just, as Belloc's confrère Chesterton had said, "the democracy of the dead"—it is also, sometimes, the potent weapon of the living.

Typing the finished essay, I thought it was the best thing I had written. An artist friend once told me that everything she painted was, on one level, a disappointment, for it invariably fell short of her original vision. I am the opposite: Almost everything I write turns out better than I imagined it would. The egotistical self-doubter. In my mind, the essay's only flaw was that it didn't carry my byline. I had come up with a pseudonym in the unlikely event that it would be published and somehow get back to the authorities who had given me a visa believing I was going to use it only to teach. The pen name I chose was Peter Kersal; it combined my admiration for the Peters I knew (like Meinke) with my current desire to have Breton ancestry (like Kerouac). The surname was more subtle than Waughbokov.

I put a few carbon copies—I had no access to a Xerox machine—into envelopes and took them to a woman I sometimes had lunch with at the American embassy, who promised to send them in the diplomatic pouch.

My teaching load increased as I took on two private pupils. One was a Japanese housewife married to a steel company executive; once a week

I would visit their house—in a small, rare neighborhood of houses—and spend an hour in dull conversation. Eventually Hania replaced me, and immediately the woman opened up, telling staggering stories of life in a strict, patriarchal society. Poland, where she had the gift of a maid, was like an ongoing vacation for her. The stories reminded Hania—not that she needed reminding—of how socially progressive her own country was, despite its economic backwardness. While I was surprised by how quickly the two of them had established a relationship that went far beyond the one I had had of student and pupil. It was an example to me of how women can connect much better than men can, a gift that helps them as travel writers.

My other pupil was an eminent cardiologist who, after our first lesson, dispensed with the textbook and talked about her life. Her past and the present were equally fascinating. I was constantly amazed, when discussing the latter, how frank she was. Though there was still a reserve, still the relationship of teacher and pupil, even when, because of her age and the nature of her revelations, the roles seemed sometimes to reverse. One of her patients was Edward Gierek, the former First Secretary, and, while she never mentioned his health, she did relay their political conversations. She was a quiet supporter of Solidarity, but—like Hania—she worried about the future. Having lived through the war, she knew how quickly idealism could be crushed.

Our lessons lasted two hours and became for me an exercise in remembering, as the steely doctor spoke almost nonstop, often disclosing tantalizing bits of information. I couldn't take notes in her presence—I didn't even dare jot things down when she went to the kitchen to brew us more tea—so my only recourse was to make mental notes. I'm sure these sessions sharpened my memory skills, another important tool in the travel writer's quiver, as I taught myself to store large chunks of conversation for short periods of time. Then, as soon as I arrived home, I recorded what I remembered in my notebook. The journal I had kept since college was quickly becoming a diary.

And it was filling with more than reported speech. I spent a lot of time in our bedroom, jotting down my daily activities and encounters, scenes from the streets and buses and churches, passages from books and articles, snippets from overheard conversations, along with news, rumors, stories, and jokes. Life had become so intense that even the mundane took on significance. Entries often ran for several pages, as I'd fatten observations with reflections, reveries, occasional rants. I was experiencing a remarkable moment in history, so the scrupulous recording of it seemed a given. Also, keeping a journal was the traditional practice of the apprenticeship writer, and the extraordinary times helped me to do it with a singular diligence.

A few days before Easter I boarded a train to Vienna. The food situation was worsening; seemingly weekly, care packages arrived at school—containing canned goods, cured meats, wheels of cheese—from churches and charities in Western Europe. (One night I literally brought home the bacon.) Since I was the only member of the household in possession of a passport, it was decided that I would spend the holiday grocery shopping in Austria.

In the morning, my overnight train ran under watchtowers along a barbed-wire corridor. There was a long stop at passport control, and then, finally, we entered a world I hadn't seen in a while: well-kempt fields—where hares hopped diagonally away from our passage—and suburban developments, the houses with sloped roofs and manicured gardens and backyard patios. The outdoor furniture looked particularly exotic. We soon stopped again and, staring out the window, I saw that the train was surrounded by soldiers. They carried guns—"Nice carbines," commented the Polish woman in my compartment—and sported jaunty red berets. Some stood at the ready, their legs apart; others walked up and down, chatting with colleagues. Later I learned that they were there to protect the trains, which frequently carried Jewish immigrants, from Palestinian terrorists.

Forty minutes later we entered Vienna under a leaden sky. I had crossed to the other side of the Iron Curtain, but I was still in Central Europe.

I found an inexpensive hotel near the station, dropped my bags in my room, and went out to explore. The city appeared to be populated by pensioners. The few young people I saw—male and female—often wore cape-like coats of Alpine green with a long center pleat in the back. Some added matching hats with a feather and a cord curlicue on the side. White stockings were popular with many of the young women; they looked fetching and vaguely colonial when peeking out from under a cape.

I liked the sausage stands—clean, efficient, and cheap. I'd dip my *wurst* into its neighboring puddle of mustard and savor the moist, fatty crackle as my teeth punctured the casing. The lost joy of street food. Then I'd sop up the remaining mustard with a slice of dark bread. In a bakery I found a loaf shaped like a baguette but with a revelatory sprinkling of salt and sesame seeds on top; it served—with a chunk of Emmental—as that day's lunch and dinner. In the cafés, waitresses brought me not a menu but one of numerous newspapers on a stick. I was used to getting my news on sticks—in the reading rooms of Warsaw—but not through such gracious attentiveness.

The supermarkets, at least in the center, were small but well stocked. I bought four grapefruits, ten oranges, a bunch of bananas (all fruits that had pretty much disappeared from Poland), two cans of asparagus, two cans of tuna fish, four cans of tomato paste, seven packaged soups, two sticks of butter, two blocks of cheese, one box of Boursin, two bottles of Maggi, a log of salami, a jar of mayonnaise, two boxes of cashews, four Raider chocolate bars, two packets of Maltesers, and three bags of chocolate-cream-filled cookies. In a pharmacy, I picked up a container of Bayer aspirin and a box of Contac. Elsewhere I added to my bounty with ten Bic pens and a roll of Scotch tape. As presents, I purchased a pair of panties for Marylka, a blouse for Elżbieta, and a blouse and body lotion for Hania. For myself, in addition to the sweets and pens, I bought a paperback copy of *Look at the Harlequins!*

Because of the weather, cold and overcast, or perhaps the materialistic nature of my visit, I experienced little of the excitement I normally felt on visiting a new city. After I had accomplished my shopping, I moped along the streets thinking of Leigh Fermor on his way to Constantinople, drawing portraits of residents to earn a little money. But he, by then, had learned to speak German.

I was so lonely I went in search of the Polish church. It stood not far from the Heroes' Monument of the Red Army and was packed for the Good Friday service. Polish Catholics were never into fellowship—Hania had been bemused to find that American churches came with kitchens—and certainly not on such a solemn day. After the service, people wandered off with their heads down, their hands in their coat pockets, their thoughts to themselves.

On Easter morning I went to Christ Church, where the service was followed—as it usually is at Anglican churches—by a coffee hour. A thick-maned man in a pinstriped suit asked me where I was from, and, when I replied that I was currently living in Poland, he said "*Jak leci?*" (How goes it?). He had worked in Warsaw for six months, he explained. He introduced himself as Herbie and then introduced his wife, Anne, who was Welsh. I assumed, from his English—and the setting—that he too was British. "No," he said, "I'm a native." Together they asked if I'd like to join them and some friends for their weekly pizza lunch. Easter pizza would be new for me, but I was eager for the company.

Herbie was what I could only dream of being: an effortless polyglot and man of the world. (It's much easier to be the latter when you are already the former.) At a young age he had learned French and English, not just speaking the languages but doing so like a native. Later he had added Spanish, Russian—which helped with his Polish—and Arabic. He was currently studying Mandarin. Like most cultured Europeans, he much preferred talking about his hobbies than his line of work, which, I eventually discovered, was finance. He was employed by an Austrian bank and had represented it in various countries, sometimes for a year or more. Herbie had everything it took to be a simultaneous interpreter, but he didn't waste it on that job.

On the drive to the restaurant he pumped me for "*anegdoty*" from Poland. Occasionally he would erupt into a spectacular cough—he was recovering from the flu—that sent his body into paroxysms behind the wheel.

At the restaurant I sat between Herbie and a friend of his who was an avid reader of travel books. This man confessed that he had enjoyed all of the chapters in *The Great Railway Bazaar* with the exception of the one about the train he had once taken: the *Trans-Siberian Express*. At first this seemed like an indictment of Theroux—calling into question the faithfulness to reality of his other chapters—but, considering it later, I realized it simply reflected the subjective nature of travel. It didn't occur to me then to ponder the future of the travel writer in a well-traveled world.

After lunch, Herbie and Anne took me to their home in a suburb, which they shared with Herbie's parents. The Russians, Herbie said, had still been around in the mid-'50s, and would sometimes knock at the door and ask to have a bath. With memories still fresh, the parents always kept the house tightly locked.

We got back in the car and drove to Kahlenberg Mountain, site of the famous Battle of Vienna. It was here, in 1683, that Polish king Jan III Sobieski defeated the Ottomans and, as the Poles had it, saved Europe for Christianity.

"So, if you think about it," Herbie said, "our present aid to Poland is simply repayment for this past ecumenical help. After so many years, with interest, it probably works out about right. Depending on how much you value Christianity."

Standing on that hill, as a chill wind whipped around us, I felt an almost patriotic pride.

Back at the house, I thanked them both for the wonderful day, and then Anne drove me into town—Herbie's cough had worsened—depositing me at a cinema that was showing a film she thought I'd enjoy. As we said our goodbyes, she presented me with a box of After Eight mints. One more treasure to take back to Warsaw.

My mother, whenever I had complained to her about not having a girlfriend, had always suggested church as a good place to meet women. Well, I thought, walking into the theater, she had been half right. Without quite realizing it at the time, I had stumbled upon a travel writer's godsend (literally), where in the future I would frequently be rewarded.

The movie was *Raging Bull*, and I enjoyed it immensely, in part because of the circumstances. In Aix I had discovered, while watching *Nashville*, the curious pleasure of seeing American movies outside the United States. Because you've been away, you notice things—like wide white belts—you haven't seen in a while; the images on the screen—and sometimes even the words—are at the same time familiar and alien. And you, as viewer, are both native and now outsider.

In addition to which there was the grand oddity of watching Robert De Niro play Jake LaMotta in the decorous heart of Europe.

The following evening, I checked out of my hotel and carried my two suitcases, heavy with foodstuffs, to the train station. The departures board displayed the names of numerous cities, two of which stood out to me: Warschau and Roma. They appeared like antonyms in a dictionary. The train to Rome was leaving before my train to Warsaw, and I stood there contemplating getting on it. Not seriously, but thoroughly. Spring would be there when I arrived. And sun. People lingering in piazzas with cones of gelato. I wouldn't understand a word, but there would be no ration cards, no standing in long lines for food. I could begin a new life, learn Italian. The women wouldn't be as beautiful as those in Warsaw, but perhaps this failing would be made up for by the stylishness of the men. I would be in a real city again, one with fashion and cuisine and color and street life.

I sullenly boarded the night train to Warschau.

Early the next morning we stopped in Brno, an industrial town conjuring cold and negativity. A trio in soiled coveralls entered my compartment and almost instantly fell asleep. The sight of them slumped in unconsciousness—not just laborers, but laborers in a wintry, communist country—worked as a cure for my self-pity.

At home my purchases were well received, making me feel a little less regret about my aborted escape to Italy.

It was good to see my students again. Our lessons focused on the spoken language and were enlivened by a back-and-forth cheerfulness. By teaching them English and being an American, I gave them a reprieve from drab routine—spending an hour with a foreigner, thinking in the language of his romanticized country—and the good feeling this produced was thrust back on me. Through snow and rain and cold, wind-swept evenings, they came looking to learn and enjoy themselves.

I was making greater progress as a language learner than they were; I had passed my beginners in competence and was now about even with my intermediate students. This was understandable, since they had English perhaps six hours a week, if they were also studying it in high school or at the university, while I had Polish every waking hour, except the ones that I spent with them and preparing for them at the American embassy. I was starting to use Polish occasionally in class—translation is always easier than explanation—though it went against the principles of the method by which I had been taught French. Demonstrating to them that I knew the Polish word or phrase for something—the first baby steps in the long journey to fluency—was definitely a way of showing off. But it sent a message that I was someone with an interest in their country and culture, which was the same as saying I was on their side. Language is never more political than in oppressed lands. I had come to their city not just to teach my language but also to learn theirs. How many people did that?

A kind of spring arrived, appearing for a day and then disappearing for a week. Leaving the apartment in the morning, I had no idea what clothes to wear and, of course, no car in which to put reserves. Some days I'd be halfway down Opaczewska when I'd turn around and go back for a heavier sweater.

No matter the weather, on weekends Jaś would propose a walk in the woods. My vote was always for a stroll in the Old Town. He craved nature, quiet, a place free of socialism; I wanted buildings, crowds, the hope of chance encounters. Sometimes I won, and as we walked down streets lined with churches and palaces, I thought of friends back home driving past billboards and used car lots. It gave me a feeling of privilege almost equal to that of being an eyewitness. The Old Town, despite its newness, contained all the features that would appeal to a wannabe European: cobblestone streets, arched doorways, horse-drawn carriages, which seemed less decorative in a city where one saw the occasional horse-drawn wagon. Yet, walking across Plac Zamkowy, I never experienced the "fullness of joy at so much life" that Charles Lamb had felt standing in the Strand.

Rain would often send us into a café stocked with cakes—so that's where all the sugar is going!—and filled with smoke. The four of us—Jaś, Elżbieta, Hania, and I—were part of the small segment of the population not addicted to nicotine. Hania's aunt Bocia, a true aunt who visited us regularly, knew to bring her own cigarettes, which she stuck with shaking fingers into an ancient, yellowed holder. At school there were two teachers' rooms—one for smokers, one for non—and you had to pass through the first to get to the second, inhaling a cloud of secondhand smoke. Of course, nobody called it "secondhand"—it was too ubiquitous, like an essential element—and nobody mentioned its potential dangers, not in a country that had seen, and might soon see again, wars, invasions, occupations, insurrections. What was a little smoke after the devastations of the last two hundred years?

So we would sip our tea in the thick haze and then walk out into the damp with hair, sweaters, and coats all smelling like they belonged to a three-pack-a-day smoker.

On one of my evenings off that spring I went to another documentary about Solidarity, this time with a colleague, Anna (pronounce both n's; every letter in a Polish word, unlike in an English one, has a reason to be there) and her husband, Jan, whose thick mustache gave him a rugged, hetman look. Jan had just returned from a work trip to South America, his first time in my hemisphere, and one of the things he had discovered was that many South Americans viewed the United States the way the citizens of most Warsaw Pact countries viewed the Soviet Union: as the oppressor.

I had never been to South America, but I had known South Americans, so this wasn't the news to me that it was to Jan. And the equating of a democratic country with an autocratic one was absurd. But the fact remained that this citizen of a nation that held my own in high esteem, that cherished it as an invaluable beacon of freedom, had gone out into the world and seen it from another perspective, in a much less flattering light. Jan had learned an inconvenient truth—at least for me, the current beneficiary of that esteem—and I was tempted to request he keep it to himself.

May Day arrived, and I went alone to Plac Zwycięstwa. Two young men nearly seven feet tall carried red flags at the front of the parade. Their presence seemed to add to the freak-show quality of the communist holiday in the land of Solidarity. Behind them marched various dignitaries, including Stanisław Kania, First Secretary of the Communist Party, and General Wojciech Jaruzelski. It was daunting to view these men at such close range, not just because I had seen them only on TV but because I had heard so many vilifications of them. Their postures and gaits—stiff and unassertive—seemed to suggest an awareness of the public disdain. An American woman, exiting the lobby of the Hotel Victoria, asked ungrammatically, "Where are they all going to?"

"Hell," replied a Polish passerby.

Bringing up the rear of the procession were Mary and Paul, the two Colorado socialists from my old Polish class.

A week later I received a letter from my mother that contained a letter from Robley Wilson Jr., editor of the *North American Review*, accepting my essay on the Tatras.

It was my first success as a freelance writer, after years of rejections, and I was over the moon, though I knew that topicality had been on my side. The quality of my writing hadn't dramatically improved, and the place I was writing about hadn't really changed; it was simply that now the world's eyes were turned toward it.

Going to school became even more enjoyable. Peter De Vries, whose novels I had read one after the other, claimed that when a man had an affair it improved his marriage because the excitement of it put him in a constantly good mood. It was only when the affair ended that he grew despondent and irritated with his wife. I wasn't sure about this theory when applied to conjugal life, but experiencing success with my writing helped with my teaching. Having a second gig, something on the side, not only made me happy, it took a little of the pressure off. And, as with the husband-turned-Lothario, it gave me another, secret identity. I was not just a teacher; I was also a writer.

Now I could say it, at least to myself.

On May 13th, the pope was shot. The news reached me on the balcony of the teachers' room before my last class. An hour and a half later, riding the bus down Grójecka Street, I saw people filing into the Church of the Immaculate Conception of the Blessed Virgin Mary. At home, Hania and Marylka and Elżbieta sat staring at the television, which showed footage of the pope greeting pilgrims in St. Peter's Square minutes before the shooting, reactions from people in the crowd, an interview with the surgeon who had operated on the pontiff, even a clip from a past trial in Turkey of the would-be assassin. There were scenes from the pope's visit to Poland two years earlier, a report from his hometown of Wadowice, and a montage of churches around the capital filled with praying citizens. We were taken to a studio in Krakow where two childhood friends of the pope were reminiscing with a local priest. As surprising as the in-depth coverage of a religious personage in a socialist country was the deft technical coordination in a land of fundamental shortages. From the news this night, Poland seemed like a normal, if eternally bedeviled, country. The anchorman read the names of leaders who had sent telegrams of concern—Reagan, Wałęsa, Mitterrand, Kania—and, signing off, said he hoped the pope would recover and return to "his great mission."

The pope did recover, but Cardinal Wyszyński died. In the evening, people once again climbed the steps of the Church of the Immaculate Conception of the Blessed Virgin Mary, this time between two Polish flags tied with black mourning ribbons. On television, the anchorman announced, "On the 28th of May, 1981, Cardinal Stefan Wyszyński went to God."

"Did you hear that?" Hania asked. "I can't believe I'm watching Polish TV."

The government declared three days of national mourning.

Schools remained open, and many students came, despite having multiple reasons not to: political activism, emotional burnout, survival fatigue. After class one evening a freshman at the university told me she had a *kac moralny* (moral hangover). A few weeks earlier, a high-school student had announced that she would miss the next class because she would be attending a mass on the anniversary of Józef Piłsudski's death.

I was noticing a growing disconnect between the rote optimism of Western observers and the beaten-down psyches of the observed. I would sit in the embassy library reading eloquent editorials on the admirable triumph of the Poles and then walk outside past crowds of the "triumphant" waiting for their exit visas.

I gave the final exams. No beleaguered students asked for exemptions; instead, on the last day, they brought me flowers.

Once again, an empty summer stretched before me.

Our first trip was to the lakes, with Jaś and Elżbieta and Hania's friend Kasia and her husband Krzysztof. Money somehow never seemed to be an important factor in Poles' warm relationship with vacation. Of course, we'd be camping, which would cut down on the cost—and in my case the pleasure. Jaś was an experienced camper; I was a recovering one. The summer after my sophomore year in college I had driven with a friend to Assateague Island, where we pitched our tent while being attacked by sandflies. At dusk their numbers were replaced by mosquitos. All weekend long there was a steady, twenty-four-hour assault as the two species feasted on our flesh in shifts. It had not made me a big fan of roughing it.

And life in Poland was hard enough; why, I wondered, increase the degree of difficulty?

It rained, of course. And the days were so cold there was no thought of swimming, even for the Poles. I learned the expression "*Mamy łagodną zimę tego lata*" (We're having a mild winter this summer). Evenings, we'd huddle together in one tent, drinking *Advocat*, which we'd bought in a dollar shop in the nearby town. Even without the alcohol, everyone except me seemed to be having a good time. I was reminded of the "winter of the century," when people had delighted in the harsh conditions.

I wasn't moved by the lakes as I had been by the mountains, in large part because I hadn't been given any signs of a culture. All I saw were trees and water (Jaś's ideal), and I missed the mark of humans. Every Sunday in Alsace, Dany would take me on a trip somewhere, and I always preferred driving through tiny, half-timbered villages to, say, a visit to the Vosges. I didn't need cities, but I needed people, a condition that made me different from most Poles—that along with a number of other things. For me, the highlight of our trip to the lakes was watching the young man at the kayak rental—a gruff, shirtless teenager—kiss a girl in rapid succession on her hand, her cheeks, and her hand again, and then repeat the procedure, just as formally, on her friend.

A few weeks later my parents arrived, my mother's desire to see her youngest son overriding my father's anxiety about World War III. They brought the June issue of the *North American Review*—my story on page 4, under the heading "Foreign Correspondence"—and a tape cassette of Woody Allen's old stand-up routines that I hoped to use as a language tool in the fall—for those students who weren't enjoying my classes quite enough.

It was my parents' first trip behind the Iron Curtain, my mother's second trip to Europe. (My father had been there in 1944.) They constituted a very small group of Americans, especially during the Cold War, who, after seeing France (when I was a student), next booked Poland. I was struck by how loudly they spoke or, more accurately, how softly Poles did—the trait of an oft-occupied people; my parents were hardly loudmouths. We took a walking tour of the Old Town, given by a man in ill-fitting clothes who made no secret of his support of Solidarity. He lost my father's vote when he blithely got us caught in a rainstorm. My mother loved the soups—mushroom, potato, beet (hot *and* cold)—and took to drinking vodka, slowly, and prefacing the first sip with "*na zdrowie.*" My father was less enthusiastic, tacitly heaping the same mild disapproval he felt for Hania on her homeland. I showed them the English Language College, pointing up to the bay window of one of my classrooms. Walking the circumference of Plac Zbawiciela, we were stopped halfway around by a nicely dressed pensioner who announced in clear English, "We are hungry." It was the most dramatic moment of their trip, though a bit of an exaggeration. That evening they came to our apartment for an ample dinner cooked by Marylka and attended by Krzysztof, who took them back to the Forum in his tiny Polish Fiat. Possibly the second most dramatic experience of their trip.

They didn't pick up on, I'm quite sure, any change in my relationship with Hania, and they were not the type to pry, or even hint, about grandchildren.

Our plan was to rent a car and drive to the south, but summer had brought another shortage; long lines of cars now waited for gas, just like those of people for food, except that the humans didn't disrupt traffic. Hania sprang into action, as she had on the day of the pope's mass, and, calling around, got us a driver. The otherwise exorbitant fee would be made quite reasonable by paying him in dollars—saved by the black market!—and he could wait in lines for gas while we toured, ate, or, as sometimes happened, slept. It turned out to be the perfect solution and should have earned Hania a place in my father's affections, for it saved him not only from sitting in queues but also from jousting with Polish drivers on the roads.

Krakow, the one major Polish city not badly damaged in the war, was an architectural masterpiece blackened with soot, the result of coal-burning furnaces and nearby mills. The narrow streets, and the buildings on them, were authentically old, but the facades in Warsaw were cleaner, and the air was healthier. Eighteen months in the capital had made me a Warsaw chauvinist.

At Wieliczka, the famous salt mine, the three of us shared a tiny elevator with a stubbled miner in a battered hard hat who was doing double duty as a tour guide.

"Do you speak English?" Hania asked the man doubtfully.

"Yeah, sure," he replied, with an almost New York accent, before explaining he'd learned it by listening to American records. His commentary was nearly as impressive as the statues and chambers carved out of salt.

Then we made the grim trip to Auschwitz. Hania wondered if it didn't all seem unreal to my middle-class American parents, while they, having lived through that time, quite possibly wondered the same with regard to us. As the one with no geographic or temporal connection, I found the personal items—the stacks of suitcases, the piles of eyeglasses—made it seem all the more surreal.

In Zakopane we stayed in the most expensive hotel, set on a hillside outside of town. This luxury, plus the bright summer light—the sun had finally appeared—deprived the Tatras of their hibernal mystique. But

I felt a proud satisfaction revisiting a place I had written about—and showing it to my parents, who had read what I'd written.

On the way back to Warsaw we passed through rolling farmland, swelled fields of wheat and barley that surprised my mother: If the countryside was lush, why weren't the shops full? Hania translated her puzzlement to our driver, who, through nods and shrugs, conveyed his own frustration with the conundrum.

A few weeks after my parents left, I boarded a train to Paris, where Hania would soon join me for a belated honeymoon in Brittany. I stayed with Mark and Sandy, who were renting an apartment on the Île Saint-Louis for what seemed to me then the exorbitant sum of $900 a month. It was on the top floor of a building that overlooked the Seine near the southwestern tip of the island; at night, the floodlights of the Bateaux-Mouches crept through the rooms. To a Francophone lover of cities, it seemed the most desirable address in the world. In the evening, walking the dog with Mark along the cobblestone quay, I became star-struck at my former colleague's new life.

I had brought my copy of the *North American Review* to show him, not mentioning that I was Peter Kersal but assuming he would guess. A few days later he gave it back to me, shaking his head with dismay at my effort. One of the things that troubled him about the story was my use of "transmogrification" in the lead paragraph. Even in a literary journal, apparently, one was not supposed to use big words. I wondered if he had stayed with the piece long enough to come across "prothalamion."

But he was far more concerned about an article he had recently read in the *New Yorker* on a movement called "the religious right." He worried that it would have serious implications for the future of our country. Having just come from one where the church was a positive force for change, I naturally thought he was overreacting.

I went to see Stefan at the Librairie Polonaise. He showed me the picture on the front page of *Le Figaro*—a long line of buses blocking traffic in Warsaw—and declared that this was the beginning of the end. Nothing like this had happened in the last thirty years. Poland was on its way to freedom. I shared his excitement and hoped I'd be back in time for the celebration.

That evening I met Hania, whether at the airport or the train station I no longer remember, but I distinctly recall emerging from the subway at Saint-Michel–Notre-Dame and the look on her face as she took in the lights, the crowds, the smells (gyros and crêpes), the sounds (underscored by a street accordion, surely)—the gorgeously intertwined pieces of a living, jostling, late-night metropolis.

We spent a few days at Mark and Sandy's, caught up with Stefan and Zbyszek, and then hit the road. We were not the only people hitching a ride out of Paris; we had to walk about a hundred yards along the highway to take our place in the queue. I had hitchhiked quite a bit in college, sometimes with a male friend, never with a woman. In France, I figured, it could only help.

It did. We entered Brittany before dark and landed in an old stone town with a picture-postcard harbor. A small hotel on the waterfront had a spare room. It was the perfect lodging for young lovers but, regrettably, we were not (back) there yet. All the charm of the place seemed somehow wasted.

The next morning, we made our way to our final destination. I had visited Brittany a few years earlier and had heard that Quiberon was where you had the best chance of finding sun. We presented ourselves at the campsite, where the manager greeted us with the news that it was full.

Hania, well acquainted with refusals from authorities and ways to get around them, explained in her most plaintive French that we'd come all the way from Poland. I'm not sure if she added that it was the land of *Solidarité*, but she might have. The man said cheerfully he'd find us a spot.

The week almost succeeded in selling me on camping. The money we saved on hotels we spent in restaurants, feasting on crêpes for lunch—washed down with small bowls of cold *cidre brut*—and, for dinner, seafood fresh from the ocean. Then we'd stroll the main street, admiring the well-dressed children *en vacances*. One day we took the ferry to Belle Île, where I bought a sweater with horizontal blue-and-white stripes and navy-blue buttons on the left shoulder. Were they there to accommodate large French craniums? Few articles of clothing could have better distinguished me from my compatriots. It seemed to go well with my mustache.

It was a pleasant if unromantic honeymoon that nevertheless revealed to me the difficulty of travel—real travel—with a companion. On my first visit to Brittany I had been by myself, less interested in fine dining—though food is an important part of a culture, one that our sybaritic age has given great prominence—and more focused on the customs and the people. I learned that "Ker" is a common prefix in Breton names (the more literate of the population claimed Kerouac as their own) and was told with pride about the high school student who had written his entire baccalaureate exam in Breton. I attended a music festival—bands from Scotland, Ireland, Spain, and Canada—that seemed to stir something in my heavily diluted Celtic blood. There was, too, the undeniable attraction, for a foreigner, of being in France with people who also saw themselves as outsiders. I returned home with an Alan Stivell cassette and a feeling of connection that this second visit, while not diminishing, did nothing to strengthen.

From France we took a train to Amsterdam, curious to see the country of playful ambassadors. Thanks to some guidebook research in Paris, I had booked a room in a private home—a pre-Airbnb. The owner, a middle-aged woman, lived with her parrot, a cockatoo that Hania took an immediate interest in. One day the woman came home from work and went straight to the kitchen, without greeting her pet. The bird grew furious, shrieking and squawking and pacing violently back and forth on the bar in his cage. When his mistress finally appeared, he turned his back on her. The woman began talking to him in Dutch, softly, soothingly, tenderly. Eventually he turned around, but he didn't move closer. She continued her loving apologies, in babyish tones, and slowly opened the door of his cage. The bird moved over, latched on to her finger, and allowed himself to be lifted out. All the while the woman continued her dulcet monologue. Assuaged, the bird started caressing his mistress's face with his powerful beak, using the sharp tool as an instrument of affection. She closed her eyes and we watched in shock as the bird delicately stroked the skin around the right one. All was forgiven.

Over the next few days we barged the canals, strolled the red-light district, visited the Rijksmuseum, toured the Heineken brewery (every tourist knew there were free samples at the end), but nothing compared to the spectacle of that parrot caressing our hostess's brow. The prize was in the home life, not in the sights.

We took the train to Hannover, where we were to catch the Leningrad Express. Finding it was easy: It was the train out of which poorly dressed passengers and an air of chaos spilled. Our time in the orderly West was over. Somewhere there were seats for us, hopefully, but I abdicated any responsibility for finding them. This was Hania's world now, these were her people, and I was leaving it to her to sort things out. I was resigned because I held her accountable: If it hadn't been for her—to use a well-practiced construction—I wouldn't have been in this predicament.

We eventually found our seats, the train departed, I calmed down. In the middle of the night—just as on the Iron Curtain Local—we stopped at a border. East German border guards stomped down the corridor with fierce-looking German shepherds on leashes. "*Taki piękny pies!*" (What a beautiful dog!), the drowsy Poles cooed. "*Wspaniały piesek*" (Wonderful doggy). Suddenly it didn't seem so bad to be on a train bound for Warsaw.

There was no uprising in the city; no joyous revolutionaries danced in the streets. The same old queues stretched with the same disgruntled citizens.

Ration cards had been extended to include vodka and cigarettes. Making our way to Moda Polska one afternoon, Hania and I found everyone on the sidewalk heading to the entrance. Inside, all the racks, save one, were bare, and a covetous crowd surrounded the two remaining sport coats.

Życie Warszawy reported that *Pravda* had called the recent Solidarity Congress in Gdansk an "anti-socialist bacchanalia." The *Herald Tribune* translated the Soviet paper's description as "anti-socialist orgy." There were inarguably a lot of shortages in the Soviet Bloc, but vocabulary wasn't part of them.

And could there have been a sudden surplus of telephones? We finally got one installed at home.

At school, some of my favorite teachers were gone—a few to the States, some to other jobs in Warsaw. Krzysztof was now a translator at the Malaysian embassy, and Anna was teaching Polish in the American embassy. My students seemed happy to be back, especially Anna and Ted, beaming in the front row in anticipation of another year of American English. Ted had expressed a liking for Polish jokes—not the dark, cynical jokes Poles produced with amazing speed, and anonymity, to greet every new political situation—but the dumb mocking jokes that were the specialty of my homeland. It showed an ability to laugh at himself or an attitude of superiority over his compatriots—I was not sure which. He, like a number of the students, had a talent for turning an innocent question into a pointed reference to the regime, which invariably delighted the rest of the class. I smiled without comment but with admiration for their ability to produce zingers on their feet—in a language that was not their own. One evening, I listed for a class the four Polish foods I found hard to swallow: blood pudding, pigs' feet, lard sandwiches, and tripe. When I asked the students if they liked these dishes, Cezary said, "Recently, we like anything." It was a response as revealing of the economic crisis as it was critical, unwittingly, of my cossetted Americanness.

I could now go through the class roster fairly fluently, taking great pleasure in the names of the female students. Agnieszka. Danuta. Grażyna. Jolanta. Katarzyna. Magdalena. Urszula. Zofia. Women's last names that would have ended with "ski" in the States here closed with "ska"—a double note of femininity that was entirely justified. How could Jolanta Jankowska be anything but lovely? Especially when you pronounced the "j" as a "y" and the "w" as a "v"?

Friday evening, for everyone but the beginners, was comedy night, as I played the Woody Allen tape my parents had brought. Even if they didn't acquire a New York accent, the students learned non-textbook terms like "Eggs Benedict" and "anti-Semitic remark."

At the American embassy cafeteria there was a new guest at the table, a graduate student in political science who, like Roman, had come to write about Solidarity. David was a serious scholar, but he had a sense of humor. One day while shooting hoops—the first and only time we wormed our way onto the embassy courts—he joked that as an adolescent he had fantasized about a country where all the women were beautiful and the men were so-so, and in Poland he had found it. With his olive complexion and unruly black locks, he got plenty of stares. After she had known David for a while, Hania asked him, in her direct yet somehow endearing way, if he purposefully tried to look Jewish. "No," he replied chuckling. "My intent was to look radical, and I came out looking Hasidic."

There was almost no talk of Jews in those years. It surely existed, but not in my circles. Everyone was focused on Solidarity, some of whose most brilliant strategists—Adam Michnik, Bronisław Geremek—were Jewish. But, as Lawrence Weschler pointed out in one of his *New Yorker* articles, they were Jewish Poles, not Polish Jews. And Poles were united, at least superficially, for the time being, against the enemy. I didn't hear—unlike Woody Allen in his talking elevator—an anti-Semitic remark during the two-and-a-half years I lived in Poland; it was mainly after 1989 that the old benighted ugliness resurfaced. Experiencing Poland in the early '80s I saw Poles, in so many ways, at their best. Solidarity was not just a free trade union that had evolved into a political movement, it was also a societal ideal.

With a dictionary (*ze słownikiem*), I started reading Polish writers. Two who appealed to me were Antoni Słonimski and Julian Tuwim, Jews who had entered the Polish canon. They had been leading figures in the literary group Skamander, founded shortly after the First World War, in the newly independent Poland, with the goal of making Polish poetry less nationalistic and lofty. Which explains my attraction; they didn't depend on historical allusions to the extent that their predecessors had, making it easier for foreigners to follow along, and they didn't employ as frustratingly ornate a style. Unfortunately, poetry was out of my reach, including Tuwim's beautifully titled "*Kwiaty Polskie*" (Polish Flowers), whose second line contains the exquisite bottom-feeder *wzwyż*—meaning, appropriately, upwards. But Słonimski was also an essayist, a humanist somewhat in the style of E. B. White. I sometimes thought of the Skamander group as a Polish version of the Algonquin Round Table, complete with the drinking and the one-liners, though in a world that was far more fraught.

Reading, I would sail along for a sentence or two and then come up against a vowelless wall. While I had made my peace with the grammar, pronunciation was an ongoing battle. Nouns and verbs appeared like unwieldy contraptions—thick, airtight, impenetrable—out of which proper words might be formed with the merciful gift of an a, e, i, o, or u (y's were in ample supply). I questioned how any human mouth could render into sound these seemingly random clusters of letters, these alphabet-ending consonant clots. And I wondered about the masochism of a people who had created such an obstacle course of a language. Numbers were the worst: "Nineteen-thirty-nine" translated to "*Tysiąc dziewięćset trzydzieści dziewięć.*" Yet when spoken by anyone other than me, Polish sounded like rustled silk.

We acquired a cat: *kot*, easy enough, as was the diminutive *kotek*, though as the endearments increased so did, slightly, the difficulty: *koteczku.* It was the same with my name: "Tom" was familiarly extended to "Tomek," giving me a welcome sense of belonging, and, on rare occasions, "Tomeczku," which, if uttered by an attractive young woman, could render me speechless—which in those situations was probably for the best.

The cat, Osman, was a peach-colored Persian. He quickly made peace with the dog and took to sleeping in our bed, climbing my pillow in the morning to tap lightly on my forehead as a notification that it was time for his breakfast. His food was purchased by Bocia, who stood in lines for fish, among other things. Someone once brought us cat food from the American embassy commissary, and Osman walked up to the open can, sniffed, and walked away with an expression of disgust and disdain. Poles who tasted the rare can of root beer, procured the same way, had somewhat similar reactions. Despite being a food snob, Osman was so likable that I attributed my itching and sneezing to the carpet.

One Saturday, Hania and I made a trip to Praga and the famous Bazar Różyckiego to look for a new pair of shoes for me. We wandered the open-air aisles, eyeing pink logs of salami and Cyrillic-scripted tins of caviar and forest-green boxes of After Eight mints—all things one never saw in the shops. On one folding table we spotted a handsome, square-tipped pair of brown Salamanders that—to beautifully cap off this visit to wonderland—fit me perfectly. Not only was the black-market price much cheaper than it would have been in the West, the pleasure of finding the shoes was infinitely greater.

On rare sunny days I played tennis with Andy—on the red-clay courts of the British Club in Saska Kępa—or with Gyuri Szőnyi, a

Hungarian who had married one of my comely colleagues and found a job teaching Hungarian. Gyuri was tall and thin with a receding hairline and a narrow beard covering a long chin. We were in identical situations—foreigners married to Poles and temporarily teaching our native tongues in Warsaw—but Gyuri was a pro, at least in the latter realm; back home in Szeged, he was a professor of English literature, with a specialization in Shakespeare. He could have easily substituted for me, while it would have been impossible for me to fill in for him. He was a man of great intellect and wide-ranging interests; conversing with him after our matches was always more interesting and challenging than hitting with him on the court. Once as we were leaving his apartment he dashed to his typewriter—where the page of an essay sat upright in the carriage—and quickly typed the next sentence, which had just occurred to him.

Home life, sporting life, intellectual life—they kept one grounded while the world spiraled. The relationship between Solidarity and the government was becoming more strained, fomenting new rumors of a Soviet invasion. Strikes grew in frequency, and food lines lengthened. One week the shelves in our local *spożywczy* carried nothing but bottles of vinegar, which seemed like a bad joke on me and my early shopping shortcomings.

Thanks to the black market, exchanging my dwindling dollars with friends, I would sometimes forego lunch at the embassy and treat myself to *pieczarki z patelni* (sautéed mushrooms) at the Hotel Europejski or borsht and pierogis at the cozy restaurant next to the Syrena Theater on Litewska Street—simple meals that felt like luxuries. On special occasions we'd dine at the restaurant in the Hotel Victoria, joining the international businessmen and assorted rogues.

"During the war," Bocia said one evening, turning her beer glass so the hotel's name no longer faced her, "the Germans had this 'Victoria' written everywhere: on walls, on trams, on buses. Even now it irks me to see it."

An unscarred American, I was often reminded of happier times. And it put me in an awkward emotional realm. If I got depressed, as I frequently did, especially when thinking of my life back home, I'd question what right I had to such a mood when the people around me were much worse off. But my occasional bouts of cheerfulness seemed, for obvious reasons, just as inappropriate and damnable. Sure, it's easy for you to smile—this isn't your country.

I felt tied to it more deeply, I'm sure, than most of the correspondents, for whom Poland was primarily an assignment. But we were all, foreigners and Poles, to lesser and greater degrees, spellbound by the daily spectacle, enveloped in the unfolding drama. The difference was that the reporters had no place else they'd rather be—safe in the knowledge that they could always get out—while many Poles, even those with an unquestioned love of their country, were desperate to leave. Anna's husband, Jan, had gone ahead of the family and taken a job at the University of Maryland.

I frequently had lunch with Anna at the embassy. Working with Americans was causing her to reflect on Polish ways.

"Jan wasn't present at the birth of Kuba," she said one day, in her atypical, carrying voice. "He couldn't be. There were fifteen women all in the same room giving birth at the same time. There was hardly room for the doctors.

"And there is a totally different idea you have in America of the 'nervous father.' Jan wasn't bothered at all; he didn't even think about it. It was something the woman did."

She took a sip of her tea. "But I wouldn't have wanted Jan there. He would probably have been telling me to do it better—as he does with everything.

"Relationships are different in America," she continued. "I think the family is closer. I mean that the husband and wife are often closer—they depend on each other more than in Poland. In Poland, you have a friend you go to when something's wrong—not necessarily your spouse. I think in America you expect your spouse to support you emotionally and all.

"You say 'we' much more in English than we do in Polish. In Polish we say 'I' a lot. If somebody asks you where you were, you say, 'I was in the mountains,' even if you were there with your spouse. Because you answer for yourself and don't attach yourself to this other individual."

Hania's friend Kasia had decided to detach herself from her husband in the most extreme way—through a divorce. But, because of the housing shortage, they continued living together. When I asked Hania how they were managing, she replied, mimicking *Pravda*, "Everything's being handled in a friendly Party atmosphere."

As fall progressed there was a feeling, at least in retrospect, like that of floating down a river that inexorably picks up speed as it approaches a waterfall. We could not yet hear the roar, let alone feel the spray, but there was a growing sense that we were headed toward a crash. Hania's initial pessimism now seemed like foresight.

But the annual holiday bazaar still took place at the British ambassador's residence. I rummaged through the secondhand books and picked up *A Winter in Arabia* by Freya Stark. At home it would sit nicely next to *Winter in Moscow*.

Bill the ethnomusicologist gave a party that, Polish-style, I attended alone. It was a small gathering: two free-spirited Italian women—university students who caused me to envy our bachelor host—and a Polish American who spoke fluent Polish and English like someone who had grown up in Brooklyn. The Italians said they could understand me—teaching your own language decelerates your speech, even outside the classroom, and fine-tunes your pronunciation—but not him. He seemed a little miffed by this and asked me where I was from.

"Joisey," he said, after I had told him. "Let's go down to the bawdwalk, listen to da Boss. Let's go to Pattason, pick up some girls.

"You guys in Jersey," he continued, now as himself, "you're always eating onion dip. In the bars there's always onion dip. And you always run out of potato chips, or pretzels, and start using your fingers."

New Jersey, I almost said to him, unexpected pleasures.

On December 12th, I picked up the telephone to call New Jersey—my father was turning sixty-two—and found the line dead. Early the next morning Piotr, our across-the-street neighbor, knocked on our door and told us martial law had been declared.

Or, as it was called in Polish, *stan wojenny* (state of war).

The news on television was now being read by a man in a military uniform. Tanks and soldiers patrolled the streets. Schools were closed until after the holidays; theaters, cinemas, and concert halls were closed indefinitely, to prevent people from congregating. Telephones throughout the country had been disconnected; the borders closed to citizens. Solidarity's leaders were being interned; General Jaruzelski jokes were being minted.

The coup was coupled with a bitter cold, the worst I had felt since the winter of the century. It abetted while complementing the feeling of entrapment. On Monday morning I put on several layers before heading out; I needed to see what a city occupied by its own army looked like.

Hania urged me to be careful. Although I was not fearful riding the familiar #15 tram—its windows delicately etched with frost—I felt a little trepidation joining my curiosity. Groups of armed soldiers appeared at intersections, gathered around fires in darkened oil drums. But they were Poles, not Russians; people I could talk to, if I had to. They might be amused by my Polish, perhaps find my presence a diversion (many were country boys who'd been brought into the capital). I felt so at home in Warsaw it was hard to imagine, for once, anything bad happening to me there.

I made the rounds: Old Town, downtown, school, embassy. My foreign-service friend had returned to the States, and, when I asked if I could send a letter to my parents through the diplomatic pouch, I was told it was against protocol. The Americans' response to the national emergency was to become more strict, unlike the French, who were accepting letters from any citizens who currently found themselves in Poland. (Ineke at the Dutch embassy would end up sending mine.) A sign on the door of the restaurant in the Hotel Forum stated that it was now reserved for hotel guests only. One day soldiers rounded up students who had been occupying a university building, and I stood and watched helplessly as Dorota from my five o'clock class was pushed with her colleagues into a police van. I really was a camera, mentally capturing moments that, in the evening, I would transform into words and put in my journal.

With martial law, everything changed, including my circumstances as a writer. Before (*przed wojną*), I had merely delighted in the fact that I had a subject that most people in the West knew nothing about. (I hadn't had to worry about writing "the same old story" about the Tatras.) Now, for the first time in my life, I felt a need, an obligation, to tell the world what I was seeing.

This new urgency traveled only so far—to the pages of my notebook, the drawer of my desk. In addition to nonworking telephones, there was no mail service. And when the mail resumed, it would be heavily censored, so there would be no chance of sending anything out. I was not driven like the correspondents, who slipped their stories to foreigners on their way to the airport, tactics that would help the *New York Times'* John Darnton win a Pulitzer Prize. I was a former feature writer, biding my time, committed to themes that were non-evanescent.

People wondered if I would be leaving. For those who didn't know I was writing, it was a reasonable question. But, in addition to not wanting to pass up on one of the stories of the century—poor Roman, the political scientist, convinced that nothing was going to happen until after the holidays, had left the country the week before martial law was declared—I felt a sense of responsibility toward my students. I still had corrected homework to give back to them: letters I had asked them to write to friends abroad—real or imaginary—telling of their lives in Poland. And how fully could I have been on their side if I had bolted at the first crisis? I was not the only one living through history—we all were—and I wanted to experience it with them. Not to mention they were providing me with excellent copy; some of their letters would appear in *Unquiet Days*.

Most important, I didn't want to leave Hania, who would not be able to get out as quickly as I and who might decide not to get out at all. Although our marriage, unlike everything else in the country, was showing signs of improvement.

Also, I wanted to stay until August, which now seemed years away, so I could walk on the annual pilgrimage to Częstochowa, where the painting of the Black Madonna—the revered protectress of Poland, whose image had adorned Wałęsa's lapel—hung in the Jasna Góra monastery.

The shock of martial law and the outlawing of Solidarity—the grotesque sight of Polish tanks in the streets—had an understandably numbing effect, particularly on people who now had friends or family interned. But everyone was affected. One of my colleagues, a young mother, was in a frantic state because she didn't know when her husband would be able to get back from West Germany, where he'd been sent for work. And of course there was no way to communicate with him. "Why did Hitler lose the war?" began one of the first new jokes I heard. "Because he failed to disconnect the telephones."

I expected the jokes but not the lack of self-pity. Downstairs one evening, preparing a more Spartan meal than usual, Kasia and Kubuś told us that they were expecting their first child. I said it seemed like an awful time to start a family, and they shrugged it off—as they did most things—saying it would probably be so for a good while, so why not do it now?

Bocia also took most things in stride, as befitted the widow of a celebrated balloonist; her late husband, Zbigniew Burzyński, won the Gordon Bennett Cup in 1933 and '35. But one day she knocked on our bedroom door when Hania was out and, in measured Polish, urged me to take her niece to America. I was reminded of the boyfriend Adam, on my arrival, instructing me to look after Hania. It was another selfless act from someone who loved her.

My students, whom I finally saw in the new year, were making the best of a disastrous situation, meeting it with stoicism leavened with humor and honed, in some cases, by clandestine activity. I was witnessing something that Hania had always told me: Poles are good in emergencies.

But it was a long winter. When the movie theaters reopened, the one closest to us showed *Picnic at Hanging Rock*, a blinding jolt of white dresses and sunshine. It was hard to imagine such a world had existed. One night on the bus home from school I ran into Kubuś and asked what he had been doing so late. He began to talk about some matter in his department and then summed it up simply as "*coś nieprzyjemnego*" (something unpleasant). I appreciated that he spoke to me in Polish, and I liked the way the genitive case ending that lengthened the adjective—*nieprzyjemny*—seemed to deepen the sense of unpleasantness.

Most nights I'd make the long walk alone, down an even more quiescent Opaczewska Street. I would arrive at the apartment about an hour before curfew and find Osman waiting at the door. In the kitchen, the women would be drinking tea, Hania already in her faded lavender pajamas with the lavender-and-white top. Marylka would raise her heavy body from her little wooden stool and shuffle over to the stove to start my omelet, while I pulled a bottle of Mazowszanka from the fridge. Cold air, especially after a night of teaching, always made me thirsty, and mineral water had more of a cleansing effect than tap water, which one wasn't supposed to drink anyway. In a land of tea drinkers, a desire for cold water on winter nights was considered eccentric. My omelet arrived plain, though occasionally, if someone had been lucky in the shops, with the cheese that went by the uncomplicated—except by diacritics—name of *żółty* (yellow). It never came with ham. *Kabanosy* had not been seen since Trenton. Matzo crackers helped neutralize the oleaginous egg.

On nights I didn't teach I'd sometimes walk across the street and play chess with Piotr. If there had been little on TV before, there was even less now. The news was still being delivered "in uniform."

On days I didn't teach I'd read: Polish at home, English at the American embassy library or the British Council. At the latter I discovered *Abroad: British Literary Traveling between the Wars*, an authoritative and entertaining study of travel writing's heyday in the '20s and '30s. The author, an American academic by the name of Paul Fussell, discussed writers who were familiar to a former English major, like Waugh and Greene, and others who weren't, like Robert Byron and Norman Douglas. I also found a review of a new book: *Lourdes: A Modern Pilgrimage* by Patrick Marnham, a writer I knew from the pages of the *Spectator*. The book covered the history of the famous French town and explored the nature of faith, but it also included an account of Marnham's journey with a group of English pilgrims.

A few days after reading the review, I mentioned the book to a new teacher at the school, a recent graduate of the university Anglistyka department whom I had told about my plan to walk to Częstochowa in August. I wanted to get the book, I said, because it would be helpful to me in writing about my pilgrimage. She laughed. "You think you can learn how to write about a pilgrimage?" It did seem a bit absurd, the way she put it, making it sound like a how-to project, something only an American, with his confident conviction of betterment, would come up with. But most novelists, with some genius exceptions, learned to write novels by reading novels, just as most poets learned to write poetry by reading poetry. I, as I've written, never took a travel writing course; I learned to write about travel by reading travel books. I wonder what she would have made of the phenomenon of writing programs.

One morning in March I walked into a downtown office and booked passage for September on the *Stefan Batory*, which traveled between Gdynia and Montreal. In 1975 I had sailed to France on the *QE2*, a voyage made possible by a short-lived youth fare, and enjoyed the experience so much that, one year later, I sailed home on the *Mikhail Lermontov*. The Poles and the Russians found ocean liners financially viable, at least for a time, because they brought in hard (Western) currency. I found them personally appealing on a number of levels. They were a throwback to a more gracious and soon-to-be-vanished era; they gave one a true sense of distance traveled; and they provided time for contemplation, which is exactly what people returning from, or heading to, long sojourns require. It had seemed unacceptable to fly home in eight hours after twelve months in France; to do the same after two tumultuous—in every sense—years in Poland was almost unthinkable. Hania, who easily got seasick, would join me in the spring.

The ticket was concrete proof that my days in Poland were numbered and was perhaps responsible for my finally tackling *Pan Tadeusz*. The towering work of Polish literature—the beloved national epic that is an ardent, unabashed hymn to Polishness—was written by the man whose name I first came across on the backs of Hania's letters: Adam Mickiewicz. He was born in 1798 in Lithuania, which was then part of the Russian Empire but had previously been the second fiddle in the Polish-Lithuanian Commonwealth. In the kind of irony not uncommon to this part of the world, the great Romantic Polish poet—the man who is to his country's literature a bit what Chopin is to its music—was born a Russian citizen in Lithuania. But he grew up speaking Polish, and he wrote in Polish; he fought for Polish independence, an act that got him exiled to the Russian interior. He eventually landed in Paris, teaching Slavic literature and pining, in poems and essays, for Poland (Lithuania). The first line of *Pan Tadeusz*—*Litwo! Ojczyzno moja! ty jestes jak zdrowie* (Lithuania! My fatherland! you are like health)—is as familiar to Poles as "Give me liberty, or give me death!" is to Americans. There was a story, probably apocryphal but nevertheless telling, that in the first week of martial law authorities went around to offices and asked secretaries to type a few words so they could have a record of the different typewriters in use, supposedly for tracking subversive materials, and every one of them, without any orchestrated plan, typed *Litwo! Ojczyzno moja! ty jestes jak zdrowie.*

After that first line, I proceeded with trepidation. As hinted earlier, I have often shied away from "great books." In literature, as in geography, I generally prefer the overlooked—or at least the somewhat less glorified. (I had made an obvious exception for Nabokov.) *Pan Tadeusz*, even with Kenneth R. Mackenzie's translation by its side, might be instructive, I thought, but not a lot of fun.

But who was having fun in Poland?

To my surprise, I enjoyed the work immensely. It contained intrigues and feuds and a problematic love story, set against the backdrop of the Napoleonic Wars, but what appealed to me were the digressive flights, the homesick homages to everything from cold borsht to singing frogs. For someone who had chosen, not once but twice, to come to this cursed land, it was like a tonic to find it being rhapsodized.

> . . . I know
> the kinds that in the south and east do grow
> And in that land of Italy so fair.
> But which of them with our trees can compare?
> . . . that much vaunted cypress, tall and thin
> To boredom rather than to grief akin?

> Is not our honest birch a fairer one,
> That's like a peasant weeping for her son,
> Or widow for her husband, as she stands,
> Hair streaming to the ground, and wrings her hands,
> Her silent form than sobs more eloquent?

I committed these lines to memory. To proclaim a tree's aesthetic superiority by showing its resemblance to mourning, its suggestion of tragedy, seemed quintessentially Polish. But this passage meant even more to me for its trash talk about Italy, the place that had so tempted me at Easter. Mickiewicz was extolling not just his homeland but also the unheralded—which I could relate to—and in so doing he made me proud that I had picked Poland. He was not just the great poet of the Poles but also the poet of those who had been drawn to Poland.

A few weeks after I finished reading the poem, Hania and I were walking with the newly divorced Kasia across Plac Zwycięstwa when we spotted a small cross someone had placed, obviously very recently, in the middle of the square. Accompanying it was a small piece of paper carrying a short, handwritten verse:

Tylko pod krzyżem
Tylko pod tym znakiem
Polska jest Polską
A Polak Polakiem
    A. Mickiewicz
(Only under the cross
Only under this sign
Poland is Poland
And a Pole a Pole)

I admired the perfect fit of the sentiment, the neat way that, thanks to grammar—the poetically useful instrumental case—two words rhymed twice. I didn't see the lines, which I learned later were not Mickiewicz's, as being anti-Semitic, or even especially exclusionary. I may have been wrong, especially considering the final line, but I interpreted them then, in that climate, as a political statement, proclaiming—on the spot where the pope had celebrated mass—the eventual triumph of a religious faith (Christianity) over an atheistic ideology (communism). One Saturday I ran into Gyuri and Małgorzata at an outdoor service on Krakowskie Przedmieście and was surprised to see that the learned professor knew exactly when to genuflect and make the sign of the cross.

Easter morning I left the apartment while everyone was sleeping and took a taxi—no trams were running at that godly hour—to the cathedral for the celebration of the Resurrection. On Easter Monday, we listened with Hania's aunt and cousins to the first underground broadcast of Radio Solidarity.

Spring arrived, tentatively as always, less acknowledged than usual. The relief of having survived another winter was muted by the dolor of martial law. It was like emerging from hibernation and hearing no birdsong.

My students Anna and Ted invited Hania and me for dinner one Saturday. We arrived at their apartment, in the northern reaches of the city, and found Ted operating with a high fever. Apparently, 103 degrees was not serious enough to cut into his social life. We started the evening with vodka, then moved to red wine, to go with the *tournedos aux champignons*. A Solidarity journalist made unemployed by martial law, Ted had turned his talents to cooking—and, clearly, shopping. The meal was followed by *bimber* (moonshine) made by a friend. Around midnight, Ted walked us to the nearest taxi stand, where no taxis waited. I suggested that he go back inside and take care of his fever, to which he replied, with a look of disappointment, "You still don't understand Polish character."

One weekend, Jaś took me out in his Syrena and tried to teach me to drive using a stick shift. I had known only automatics, and every car in Warsaw had a manual transmission. Though most were easier to operate than a Syrena, which I never succeeded in getting out of the parking lot where we had our lesson.

From the passenger seat, I watched in admiration as Andy instinctively shifted gears. We were driving south on Puławska to visit his friend, a former flight attendant who now served as secretary to the Swedish ambassador.

Ingrid stretched on her chaise next to a hardcover copy of *Out of Africa*. I had never heard of Isak Dinesen and was intrigued by the fact, relayed by Ingrid, that she was a Dane who had written about her farm in Kenya. Like Stark in Arabia, Durrell in Greece, M. F. K. Fisher in France, Gerald Brenan in Spain, she was a domestic travel writer, someone who had settled in another country, discovered its practices and prejudices, noted its rhythms and textures through the seasons. It was what I was doing, only I hadn't followed the sun; I had followed a woman. To my advantage, the news had followed me.

As Ingrid served us tea and cookies, her daughter appeared, a pretty eight-year-old with long blond hair falling onto a black sweatshirt that carried the words "American School of Warsaw." She walked up to me, put out her hand, and said, "I'm Isabella." Then, after greeting Andy, she looked at her mother and said, "Mom, we have to make a sentence using the phrase 'before God.'"

"Well," Ingrid said, after giving it some thought, "you could say, 'I feel small before God.'"

"But I don't know if I believe that there is a God."

"I thought you told me you did believe."

"Sometimes I do and sometimes I don't."

"Well," said Ingrid, "you better tomorrow."

Before leaving, I asked if I could use the bathroom; I loved roomy, unrigged, diplomatic corps bathrooms. Ingrid's came with a scale. Standing on it, I learned that I now weighed 59 kilos. It made no impression on me until I got home and did the math: 130 pounds.

The morning of May 1st, as I dressed to go out, Hania once again urged me to be careful. A service was going to be held in the cathedral, after which a protest march was planned, as a counter–May Day parade.

On the street I passed a worker, a tragic figure in a comic book outfit: a soiled cloth cap; an ill-fitting gray suitcoat, shiny with age; purple bell-bottom trousers. He was walking away from the buses that would have transported him to the parade in his honor.

Mine took me past tourist buses filled with policemen. I jumped off at the corner of Jerozolimskie and Nowy Świat and heard my name. Turning, I found Jeremiusz, a former student enrolled in music school. I asked how he was faring in the "state of war."

"I have a wonderful profession," he said. "I can forget about everything else."

The next bus dropped me in front of St. Anne's Church on Plac Zamkowy. The square was unusually active for nine thirty on a Saturday morning and was touched by an unspoken "something" in the air. People walked in looping, silent circles. A matronly woman strolled on the arm of her husband, a small Solidarność pin criminalizing her lapel.

Buses deposited more and more people, including another former student, Marek, who wore a pin of the Black Madonna set against a backdrop of the Polish flag.

After twenty minutes, a great movement of bodies began across the square in the direction of Świętojańska Street. We followed it past storefronts and into the cathedral, where people were already packed into pews. The sound of clapping reached us from outside. Seconds later, a man entered the church holding high a Polish flag with the word Solidarność written across the top white half. A foreign television crew bathed it in a brilliant light. Applause now thundered through the cathedral. The shock of hearing applause in church was almost as great as that of seeing this forbidden name.

A frail, poorly dressed man entered the sanctuary carrying a hand-written sign: "*Solidarność—Była. Jest. Będzie*" (Was. Is. Will Be). He laid the sign down, fell to one knee, and made the sign of the cross. A group of schoolgirls followed, carrying a three-poled *SOLIDARNOŚĆ* banner. After the last of the flags and banners had entered, my hands were red from clapping, and my knees were trembling.

The service began. The sermon was on the subject of the worker, and mentioned the pope's recently published encyclical on work, which he viewed as the means by which human beings help one another. The priest went on to describe the lamentable condition of the Polish worker—without motivation, unable to find meaning, knowing that promotions are given to members of "privileged groups."

Like everyone (except me), the girls with the banner lined up for communion. Looking at their fresh, serious, concentrated faces, I was reminded of a painting I had seen of Polish soldiers taking the wafer before heading into battle. For no one knew what awaited outside. The reemergence of the Solidarity name had had a jarring effect on me; what, I wondered, would it do to the police?

The service ended with the patriotic hymn "*Boże, coś Polskę*" (untranslatable, but often rendered as "God Save Poland"), which people sang while making the V sign with their raised right hands. The woman next to me sang as tears ran down her cheeks.

Outside, the mood quickly changed; a loud, throaty, masculine roar rose up: *SO-LI-DAR-NOŚĆ! SO-LI-DAR-NOŚĆ!* The crush of the crowd was so great I lost all individual movement; I was simply carried along by the current. *SO-LI-DAR-NOŚĆ!* It occurred to me that some of this lusty defiance might be an attempt to banish fear; packed within that narrow street, we had no means of escape—except back into the cathedral—in the event of an attack from both ends.

I was relieved when we reached Plac Zamkowy. Marek, gazing down at the stream of people still flowing out of Świętojańska Street, said with satisfaction, "And this is the great communist holiday!"

The procession, now less tightly bunched, headed up Senatorska Street. Residents leaned from open windows and were greeted with chants of "*Chodź z nami!*" (Come with us!). In front of me, mothers walked with their children, fathers carried their sons on their shoulders. Their presence seemed like a shield against force.

We made a turn onto Miodowa Street, and, when I reached the intersection, I discovered why: a row of policemen standing shoulder to shoulder in their bluish-gray jackets. Behind them sat green military trucks filled with reserves. They provoked more chants: "Free Lech!" "We want Lech!" "Lech is with us!" At one point people moved to the side of the street to avoid stepping on the freshly painted graffiti on the cobblestones: "*Solidarność wygra*" (Solidarity wins).

We soon made another right turn, to avoid a paramilitary unit holding carbines. "*ZOMO do domu!*" (ZOMO go home!), people chanted. Also, "Who are you serving?" A young mother watched from a balcony with her baby in her arms; another woman leaned out of her window and clapped rhythmically as we passed. I took in my fellow marchers, the trees green with buds, a world awakened, and thought: Prague. Warsaw. The Eastern European spring.

I lost Marek but ran into a high school student I had examined for admission into the English Language College. Agata's dark brown hair was pulled back into an abbreviated tail, and she wore a quilted black jacket and a long black skirt. White socks and colored sneakers covered her feet. I asked if she was scared.

"Yes, I am," she said.

"Do your parents know you're here?"

"No. They'd probably be worried."

The march ended peacefully along the Vistula. Four young men climbed to the roof of a construction caravan and asked for a moment of silence for the interned. Then they informed everyone of a gathering on the 3rd—Polish Constitution Day. Two police cars and an armored vehicle cruised slowly along the highway.

I walked with Agata back to the Old Town. She wondered if I had come out of curiosity, and when I told her I had, she asked piercingly, "But do you feel sympathy for the cause?" I had never seen a sixteen-year-old so intense and burdened. Not once in the twenty minutes we had walked together had she enthused about the march, gushing, as a teenager might, in mindless superlatives. At one point she said, employing that plaintive tone that seems to pass from generation to generation in Poland, "But it doesn't mean anything."

She spoke about the "absence of possibilities"—the difficulty of finding an apartment, the difficulty of finding meaningful work, the new inability to travel abroad. She had her life in front of her, a life increasingly drained of promise.

She asked about demonstrations in the States. I mentioned that, in college, I had traveled to Washington once to participate in a protest against the Vietnam War. Her eyes grew large. "An American who marched against the Vietnam War! That's a whole legend!" I explained to her that, at least when I did it, there had been little chance of violence.

I got a bus back to Ochota, making it in time for Kasia and Kubuś's second party in two weeks. Because of the small size of their apartment, they had invited half of their friends last Saturday and the other half this Saturday. One of them mentioned that he had heard rumors of a demonstration in the Old Town; Hania, who was sitting on the couch, assured him they were true and then presented me as proof. As I gave my account I noticed, out of the corner of my eye, Hania gazing up at me with an expression I hadn't seen in a while, one that seemed to reflect a mixture of admiration and pride.

Disillusioned in love, I had poured my passion into Poland, an undertaking—it was not a strategy, except professionally—that apparently was winning the love back.

The following week, in the embassy library, I read Darnton's story about May Day in the *New York Times*. He mentioned that there had been a service in the cathedral, but it was clear he hadn't been there to witness it. I felt as if I'd gotten a scoop—though one that wouldn't be published for years, a fact that technically deprived it of the designation.

As the school year headed to a close, I paid less attention to the textbook. One exercise everyone enjoyed was when I played a cassette I'd taken out of the British Council. It was not a language tape per se; it contained half a dozen tracks, each one a scenario comprising a sequence of various sounds. Students listened to the sounds and then invented a story that would explain them.

My favorite featured gently splashing water—a woman taking a bath?—followed by a record of Frank Sinatra singing the opening lines of "All or Nothing at All." It was a slow, wistful arrangement that went straight to my heart every time I played it. I had never really cared for Sinatra—he was someone my parents liked—but this piercing snippet made me want to listen to all of his songs, including the rest of this one. There was something about hearing that clear, strong, impeccably American voice in an English classroom in the middle of Warsaw that made its yearning brilliance apparent.

It became the impetus for my final month's focus on Americana. It was as much for my students' edification as for my well-aged homesickness. I took a cassette of Johnny Cash songs out of the embassy, and one of them, "Ridin' on the Cotton Belt," prompted a brief lesson on American belts: Borsht, Bible, Corn, Rust, Sun. The students were surprised to hear about the first; many no doubt longed for the last. I showed them copies of the *Herald Tribune* and the *New York Times Book Review*, with its risible list of bestsellers. I used the pictures in *Sports Illustrated* to try to explain the mysteries of baseball; they roared when, demonstrating the pitcher's motion, I employed the high kick of Juan Marechal. I brought in my copy of *New Jersey: Unexpected Pleasures* and talked about the equally exotic phenomenon of malls. I read them Thurber stories and fables—"University Days," "The Unicorn in the Garden"—and their laughter confirmed the universality of the myopic Midwesterner. One early class I held outside in the park, the nearby Dolina Szwajcarska (Swiss Valley). It was a beautiful afternoon, and it seemed an American thing to do. On our way back, we passed Mr. Kuczma in the school passageway. Later, at tea, he told me never to do that again. "If the police see a group of young people like that, you may never get back home." The teachers added their own imagined consequences.

Since I was talking about my home, I asked the students to tell me about theirs. "West of the Vistula is Europe," Michał declared; "east is Asia." This moving of Metternich's dividing line from Vienna to Warsaw got more laughs than protests. Małgorzata suggested I might want to visit a street in Praga (Asia) named Brzeska, as Marek Hłasko had written about it in one of his novels. "There is a hospital there," Michał added, "especially for people who enter that street."

Other students offered more sights. Some I knew—Muranów (site of the Warsaw Ghetto), Blikle (the venerable pastry shop on Nowy Świat), Winnie the Pooh Street (ulica Kubusia Puchatka), Joke Street—but many I didn't: the Muslim cemetery, the lone subway station (constructed in the '50s before the project was abandoned), the longest fence in Warsaw (in Mokotów and apparently serving no purpose), the little movie theater where they waited until at least five patrons showed up, and the sidewalk around the Polonia Hotel at night.

My last day was rich in flowers and emotional farewells: to Mr. Kuczma, the teachers, the students, the staff. I had recently turned thirty, and I was leaving the job I had held longer than any other. At the end of the evening I took the attendance sheets out of my folder and placed them in my suddenly light bookbag. I might over time forget most of the faces, but I would always have the names.

That Saturday, we hosted a party for the two advanced classes. It was interesting to see my students outside the classroom, in an adult, social setting—where they looked perfectly at ease, education and experience having combined to give them maturity. And they, I suspect, were just as intrigued to see their American teacher in his small Warsaw apartment with his Polish mystery wife. One of the Katarzynas, the one with sparkling green eyes and a summery smile that always managed momentarily to banish the gloom, drank a fair amount of vodka and lingered till it was too late to go home. I tucked her into Elżbieta's bed—everyone but Hania had found other quarters for the night—and then slowly, dutifully, retreated to my own.

My last carefree summer began in front of the television. The World Cup had started in Spain, my introduction to the magnitude of soccer. Initially, I was astonished simply to see what men could do with their feet. The virtuoso displays of pedi-dexterity—here enacted by the greatest players in the world—seemed to expand the concept of graceful. So this is what's meant by "The Beautiful Game." There were collisions, to be sure, but it was the field sport that most resembled a dance, one that flowed back and forth over the grass with hardly an interruption. The ball was almost always in motion, unlike in baseball, a game I love but many people find equally lacking in dramatic moments. The sudden bursts of intense activity, especially when they resulted in a score, were less frequent than at the ballpark and so, understandably, they produced more euphoric celebrations—or deeper sloughs of despair, depending on which team had done the scoring.

The other surprising aspect to me was how many non-teachers had time to watch soccer in the middle of a weekday. For decades, Poles had taken a cavalier view of work—"They pretend to pay us and we pretend to work"—and it had only become more so during martial law. Work that had seemed meaningless before, now appeared hopeless. Life wasn't far behind.

Then all of a sudden, into this despond, appeared the biggest sporting event in the world. And Poland, a country that had literally been cut off from the world, was a part of it. For the last six months there had been little on television and nothing to cheer about. Now the masses had their beautiful opiate.

Poland's national team, like those of many countries behind the Iron Curtain, was strong and talented (I was told), the result of a system that put great emphasis on sports and saw competition with the West as an opportunity to demonstrate superiority. Though, for the Poles who cheered the men in white and red, every goal they scored was a goal for Poland, not an ideology.

As the tournament progressed, I saw how soccer can transfix a nation and lift its spirits. For four weeks Poles focused their attention on something other than politics; the name "Boniek" (the star midfielder) was heard more often than "Wałęsa." Admittedly, Boniek was having a better summer.

My favorite player was the veteran Grzegorz Lato, whose short, unbarnacled surname was the Polish word for "summer."

Sports, of course, are never played in a vacuum. During one early-round match, an errant shot by a Polish player revealed fans in the stands holding Solidarity banners. Roars of delight spilled from open windows. It was suggested, among the group I was watching with, that the player had intentionally missed, so to force the camera to linger on the signs with the familiar, jumbled red font. A few more shots on goal, or banners, followed until, eventually, the Polish station that was receiving the feed from a Spanish one began splicing in a random crowd scene whenever the ball sailed in the direction of the demonstrators. The wide world of censorship.

Poland made it to the semifinals, edging out the USSR, a satisfying achievement in any season but especially that of martial law, which most Poles believed had been ordered from Moscow.

Immediately after the tournament I started training for the pilgrimage, which was now less than a month away. Happily, the weather was uncharacteristically warm and sunny. One morning I set off down Opaczewska and kept on walking to the center of town, crossed the Poniatowski Bridge—my first time over the Vistula on foot—and strolled the length of Saska Kępa all the way down to the British Club, where I played two sets of tennis with Andy. Another morning I walked to Wola—where Phyllis, the American Mennonite teacher, served me a breakfast of sour yogurt and tea—before heading up to Muranów. I skimmed the parks—Łazienki, Pole Mokotowskie, Ogród Saski—but never entered them. On their quiet, tree-lined lanes I would have felt as if I were in any European city, while here on the sidewalks, as I dodged the queues and the imperiously parked cars, eyed the drab mannequins and the stylish young women, there was no doubt as to where I was. I had developed extremely warm feelings for Warsaw—it was not just the city I had lived in the longest, it was the one I had been through the most with—and these long walks cemented my attachment. It lacked the ethnic neighborhoods and rich human gallimaufry I was used to finding at home, but it had its share of characters; they were just all *Polish* characters or (see Mickiewicz) Lithuanian, Ukrainian, Jewish, Tatar—hence, the beauty that went far beyond blond. Its lowly status—not just among capitals, but vis-à-vis Krakow—its shortage of glamour, its grim and doughty history had all, over time, endeared it to me. As had, of course, my learning its language. And my affection was sharpened by being uncommon. For the correspondents, Warsaw was just one city—albeit an intoxicating one—in a career of foreign postings. And expat writers owned Berlin, London, Rome, Paris. Everyone, not just the literati, had Paris—and, so the saying went, always would. Warsaw felt like mine alone. Such is the beauty of unsung places.

On rainy days, which were surprisingly rare that summer, I went to the public library just off Plac Konstytucji and read Władysław Reymont's *A Pilgrimage to Jasna Góra.* Reymont, who was awarded the Nobel Prize for Literature in 1924, walked the pilgrimage in 1894 and wrote this slim volume that, luckily for me, had been translated into French. (French, because of all the cognates, was still much easier for me to read than Polish.) I wasn't allowed to check the book out, so I sat at a desk reading and writing down salient passages that, I imagined, would be dropped into my account as I made comparisons between the contemporary pilgrimage and those of the past. It gave me a sense of purpose in an aimless if ambulatory summer.

One afternoon, riding a bus down Nowy Świat, I ran into Ted. We talked about the World Cup, and I told him about my upcoming voyage on the *Stefan Batory.* "I would love to sail on the *Batory,*" he said in a voice of almost impossible longing, and I immediately felt bad for having mentioned the ship, which now felt like gloating. A plane wouldn't have given my departure such a romantic aura or, oddly, the same enviable notion of flight.

The morning of August 5th broke clear and calm. I said goodbye to Hania and took the bus, with a group of French pilgrims who had stayed in our apartment, to Plac Zamkowy. The square overflowed with people and filled me, finally, with an urban if rebellious joy. It was the largest gathering of Poles since the imposition of martial law seven-and-a-half months earlier, and it was going to weave its way through the country for the next nine days—a moving, ever-expanding rebuke to the regime.

We walked through the city like a liberating army. The entire population, it seemed, had come out to wish us Godspeed; those stuck in offices gave us the V sign from upper-floor windows. My love for Warsaw grew even stronger because it felt as if it were being reciprocated.

The weather stayed hot and dry. We awoke in the coolness of dawn, sometimes before it, and walked all day under a constant sun. Around sunset we would arrive in a village, occasionally just a field, and find the trucks that carried our backpacks. I had not seen such fine organization in two-and-a-half years in Poland. Then we'd set up our tents, as the barns—popular in Reymont's day—were invariably full. Before bed I'd join the line of pilgrims waiting by a well; the popularity of the pilgrimage, especially this year's, had brought the city's queues to the country. When my turn came, I'd wash my feet and brush my teeth.

The monotony of putting one foot out in front of the other, hour after hour, day after day, was relieved by the daily programs—a series of prayers, sermons, hymns, and discussions. Pilgrims were divided into groups, and each group had a microphone and loudspeakers. The voices gave you something to focus on beyond your own physical effort or bodily pain—religion in a nutshell—though stragglers fell to the back and enjoyed more private conversations. I listened carefully to the discussions—which, while based in theology and often focused on morality, invariably veered into politics—and I was moved by the patriotic songs. The rosary and the chanted prayers, like the *Godzinki*, created a repetitive rhythm that proved extremely conducive to walking.

During breaks I quietly filled the small notebook I had bought for the pilgrimage.

There was, for me, the pleasure of seeing the countryside after years of living in the capital; entering small towns, whose residents always came out to greet us, and thatched-roof villages of mud-caked cows. It astonished me that the language I had learned in Warsaw also worked here, in these blunted backwaters. And I enjoyed discovering Hania's country without Hania, making my way in Polish without her help. There was no better way I could have concluded my time in Poland.

After nine days and nearly 140 miles, we entered Częstochowa. The streets were lined with well-wishers, though the mood was more subdued than it had been in Warsaw; the faces of the old were gaunt and doleful. When our group finally made its way up to the monastery and into the chapel, I remained more observer than supplicant. I was surprised by the crush of bodies, the forceful jockeying for position, the human element in a holy place. But it was evidence of the Black Madonna's power over Poles, the phenomenon of a sacred icon—that painted expression of mournfulness and hurt—serving not just as a solace but also a mirror to the nation.

On the train back to Warsaw I stood by an open window in the corridor. It was a warm August evening, and I was in a contemplative mood. I had just taken part in a momentous event at the end of a convulsive and enriching two years. The land that passed—ridiculously fast, without my having to move a muscle—was one that I had now walked across and slept upon and, most mornings, fertilized. I had waded in its streams and trudged through its forests. I had inhaled its dust and drunk from its wells. I had marveled at beanstalks. I had seen farmers who resembled peasants and village girls dressed like folkloric dolls. I had listened to a footsore actor recite lines from Mickiewicz's *Ode to Youth*. I had heard—I had said—prayers for the poor, the sick, the interned, the nation. I had attended evening services. Entering Częstochowa, I had been handed a bottle of soda by a wide-eyed child. I had knelt before the Black Madonna. I had learned stirring anthems like *Rota* (The Oath) and *Modlitwa o Zwycięstwo* (A Prayer for Victory), the words of which I was still singing as the train pulled into Warsaw's Central Station:

Krwi nie wołamy, zdobyczy nie chcemy,
Nie chcemy mordów, do łupiestw niezdolni,
Tylko odzyskać Ojczyznę pragniemy,
Tylko być wolni

(We don't ask for blood, to conquer is not our plea,
We don't want killing, for pillage we have no gift,
We want only to regain our Fatherland,
Only to be free.)

I began the half-forgotten business of packing. In addition to my clothes and books—my long city walks had often been interrupted by stops at *antykwariaty*—there were my wooden figures and a tableware set Hania had purchased. They filled a dozen boxes that revealed to me another advantage of traveling by ship. My journals—the two loose-leaf notebooks that contained my writings from the last two years—and the small pilgrimage diary could not be included because, with martial law in effect, everything had to pass by the censor. I didn't bother approaching officials at the American embassy; if they declined to take a letter, they'd surely refuse to accept my notebooks. Ineke offered to put them in the Dutch diplomatic pouch and send them to her mother in the Netherlands, who would then mail them to my parents in Phillipsburg.

On a rainy autumn morning I said goodbye to Elżbieta and Marylka—who made the sign of the cross on my forehead—and Hania and I got a taxi to the Central Station. As the train neared the coast, the woman in our compartment asked where we were going. Hania told her that she was seeing me off on the *Stefan Batory*.

"You're letting your husband go on the *Batory* alone?" the woman asked critically. Apparently, onboard life was such that the ship could have been renamed *Eros*.

We took a taxi to the Grand Hotel in Sopot, our last black-market splurge. Jaś, who had driven the boxes up in his Syrena, was waiting in the lobby. The woman at the reception desk found my reservation and asked to see my passport, visa, and work contract. After looking them over, she announced that I would have to pay in dollars. Hania and I reminded her that I'd been a teacher in Warsaw and so was not subject to the rules that applied to tourists. The woman replied that, because my contract had expired in August, I was technically now a tourist.

"I worked for two years in this country," I said to her in Polish, "and now for my last two nights I must pay in dollars like a weekend tourist?! It's absurd!"

"There are a thousand and some absurdities in this country," the receptionist said calmly, "but they're not my fault. I have regulations that I must follow."

We asked if we could get something to eat.

"The restaurant is closed," she said. "The only place with food is the nightclub."

We followed the sound of execrable music down a flight of stairs, where we were hit with an exorbitant entry fee. After several minutes, a waiter reluctantly appeared at our table.

"Are there cutlets?" Jaś inquired. Often in Polish restaurants then, you didn't say what you wanted, you asked what they had.

"No," the waiter said, with what sounded, oddly, like satisfaction.

"Potatoes?"

"No."

"Cold cuts?"

"No."

"Cheese?" Jaś was now systematically going down the menu.

"No."

"Mushrooms?"

"No."

"Pickles?"

"No."

"Bread?"

"Unfortunately, yes," the waiter said before turning and shuffling back to the kitchen.

I was perplexed.

"Why 'unfortunately'?" I asked.

"Because if they didn't have bread," Jaś explained, "he could go home."

While waiting for our slices, Jaś gave a little speech. He expressed his admiration for how, from the moment of my arrival, I had been determined to learn about Poland. I thought back to that evening in the kitchen two years earlier and Jaś's incomprehensible lecture about boats. Now I was about to sail off on one, with what felt like an imprimatur.

Our bathroom in the hotel had a tile floor of large black-and-white squares. I closed the door and immediately banged my toe on a bolt that served as a doorstop. It was painted black and rose from one of the black squares. Blood now added color to the design.

The next morning we drove my boxes to Gdynia, where the yellow, red-ribboned funnel of the *Batory* rose over the port. A porter loaded everything, including us, onto a freight elevator that opened onto a large, airy room filled with counters, desks, and scales; vintage Polish travel posters hung forlornly on the walls. We announced ourselves and then sat and waited. Thirty minutes. An hour. An hour and a half. Finally, one of the customs agents motioned us over.

He told me to open all of the boxes, even the ones that had been signed and sealed by the censor in Warsaw. Hania tried to soften the man up with friendly banter.

"But you must know by now, sir, who's smuggling something out and who's not."

"You never know, ma'am."

"Ah, but sir, you must have an eye for those kinds of people."

"But I must do my job."

He pointed to one of my wooden figures, wrapped in newspaper. It was my most recent acquisition: a stout, expressionless Polish soldier holding a rifle in one hand and a grenade in the other. Hania hurriedly unwrapped it and held it up close to the inspector's nose.

"He's lovely, isn't he?" she said.

The inspector seemed unsure.

"There's so much love for the military in our country now," Jaś said wryly, "that this is an ideal souvenir."

In the afternoon, we drove to Gdansk to see the Monument to the Fallen Shipyard Workers of 1970, which had risen just months after the birth of Solidarity. We parked and walked past the gates of the Lenin Shipyard, policemen observing us from the fringes.

The monument was taller than I had imagined it, its three crosses rising as high as the nearby cranes. Nailed to each, like the body of Christ, was an anchor. Close to the ground, the base of each cross widened to accommodate biblical passages and modernistic figures of thin, fragile workers ages removed from the socialist-realism tableaux of Plac Konstytucji. Covering the pavement around the monument were fresh flowers, photographs, and pins—that in memoriam display that Poles, through practice, had developed into an art form.

Checking out the next morning, we told the receptionist of the injury to my toe and suggested that putting a bolt in a bathroom floor, and then camouflaging it perfectly, was not a great idea. She expressed about as much sympathy and concern for my pain as our waiter had for our hunger. Our stay at the Grand Hotel had been like a final test of my love for Poland. Though I knew it wasn't all Poland's fault.

We drove to the port through autumnal streets. I kissed Jaś three times on the cheek and gave Hania a long, hopeful hug. Then I entered the terminal and walked the gangway out of Poland.

One of the many marvelous aspects of a ship is the fact that it creates its own universe, one that seems blissfully divorced from life on land. This makes it—or *made* it—the ideal vehicle for difficult departures. I was a tangle of emotions, of course, but I found myself in a new, unexplored realm—with nooks and public rooms, travelers and immigrants—that not only helped take my mind off the multiple separations but also carried it back, as I prowled the decks, to previous crossings.

The *QE2* had resembled a college dorm, at least for those of us traveling on the special youth fare. I shared a porthole-less cabin in the bowels of the ship with two other recent graduates, one of whom had been assigned the same table in the dining room as I. The first evening we were joined by two charming young women who were equally new to shipboard life. Realizing that it guaranteed me a dinner date every night of the crossing, I embraced it.

I discovered the rituals—the captain's cocktail party, afternoon tea—and loved standing out on deck and seeing nothing but water in every direction. A daunting sight during the day, at night it became exhilaratingly dire, that presence—surrounding our blazing lights, there where our blazing lights had just been!—of a vast, dark, deathly emptiness.

I had no luck with our dinner companions. The friendlier of the two had recently gotten engaged back home in Missouri, where she played the flute in the St. Louis Symphony. She was headed to Europe for a series of concerts. One night we sat in one of the dimly lit lounges and I told her of my plans to learn French, have experiences, and, hopefully, become a travel writer someday. It was information I had not shared with many people; the confessions one makes to strangers come even more easily on a ship, with its self-contained population adrift in a cocoon of acutely compressed time.

Time was my enemy. I was anxious about the year ahead—living in a country where I didn't speak the language—and so, understandably, in no rush to get to it. But the *QE2* crossed in five days, barely enough time for passengers to learn its layout. And, because we were sailing eastward, every night we lost an hour.

I got them back the following year when I sailed home on the *Mikhail Lermontov*. My Nigerian friend Ivy, from my food hall days, drove me to Tilbury on a rainy evening, and I remember pitying the people trudging with umbrellas while I dashingly made my escape by sea. Not so different from my feelings that morning riding through the leaf-sprinkled streets of Gdynia.

The *Lermontov* took a week to reach New York, thanks to a stop in Le Havre. The officers, especially the stocky, unsmiling women, were stark reminders that we were on a vessel of the Soviet Union. As was the food: a predominance of borsht and boiled potatoes. Entertainment consisted of songs and dances performed on stage by our waiters and chambermaids, barely recognizable in their colorful folk costumes. This dual service, while saving the ship money, gave it an endearing, homey feel.

But the real pleasure was found in the late-night conversations. The passengers constituted a rich mix—students, Peace Corps workers, immigrants, retirees, the inevitable German backpackers heading to South

America, the proverbial American promoting peace—and most of them had interesting stories to tell. During the day, I would find a quiet spot and read a translation of Lermontov's *A Hero of Our Time*.

I recalled those voyages now, standing on the deck of the *Batory*. When the ship finally inched away from the pier, a small band below played a heart-wrenchingly slow version of the Polonaise by Oginski, "*Pożegnanie Ojczyzny*" (Farewell to the Fatherland). With centuries of practice, the Poles had also mastered the art of the send-off.

As on the *QE2*, I shared an inside cabin, though this one bore possibly the most difficult number for me to say in Polish: *dwieście trzydzieści trzy* (233). My dinner companions were a suave and shady Polish man and an amiable Swedish foreign service officer traveling with his wife to his new posting in Los Angeles. The fact that the Pole had been seated with us only increased for me his air of suspicion. Along with the sensation that I had not left Poland.

Scattered around the dining room were a good number of immigrants, including families.

We arrived in Rotterdam—its port so large we seemed to be making our way back to Poland—and then headed to Tilbury. Two years had passed since I'd been in an English-speaking country, if you don't count the Netherlands, and, just as in 1976, I went immediately to a bookstore. A paperback copy of S. J. Perelman's *The Last Laugh*, with an introduction by Paul Theroux, caught my eye. We were less than a week into our two-week crossing; it would help fill the hours while satisfying my craving for American humor and virtuoso wordsmanship. Also, there was something about sailing on a Polish ocean liner and dining every night with the newly appointed Swedish consul in Los Angeles that seemed mildly Perelmanesque.

At dinner that evening, the consul noted that the Polish children already seemed to have acquired a healthy glow, not so much from the salt air as from the steady diet of vitamin-rich meals. The Pole asked him what he thought of the Soviet submarine that had recently been spotted in Swedish waters. He replied that perhaps he should eat his meals under the table.

During the night, the ship started to roll. I unsteadily made my way to the dining room for breakfast, and, shortly after sitting down, stumbled out in the direction of the lavatory. My mouth and chin cleaned, I returned to my cabin and climbed onto my top bunk. The steward making his rounds asked what was wrong. I told him, and he insisted that I get outside, in the fresh air. He instructed me to stand in the middle of the ship and keep my eyes on the horizon.

He meant, I soon realized, the horizon in front of me. When I looked to the side, it rose and fell with a severity I had seen depicted in movies and always thought an exaggeration, a special effect. I watched with grim fascination as the stern soared high above the waves and then, cinematically, sank far below them. I had not gotten seasick on either of my previous crossings, and it somehow seemed fitting that a Polish ship was giving me the full oceangoing experience.

I was not the only stricken passenger. Stumbling into the dining room for dinner, I found it almost empty. I knew I should eat—the pilgrimage had made me even skinnier—but the sight of my soup swirling in its bowl had a similar effect on my stomach. I made it to the lavatory just in time.

The rough seas continued. Then one evening, as I sat outside the dining room wondering if I might be able to force some food down, the world all of a sudden returned to normal. I wasn't sure if it was because the ocean had calmed or my body had acclimated to the ship's movement—probably a combination of the two—but in an instant I went from feeling queasy to having a huge appetite.

The next day passengers reappeared; the seas *had* quieted. I ran into a woman I'd met before the tumult, a petite blonde by the name of Wacława (not one of my favorites). I had first spotted her walking through the streets of Tilbury in a green wool skirt and matching jacket. She was from Krakow and on her way to make a new life in Chicago, where, like many Poles, she had family. Here on the ship, she had a cabin to herself. With a porthole. I immediately wondered how she could afford it.

After dinner she took me to see it. There were two beds; we knelt on the one beneath the porthole and gazed out at the ocean. It was now calm, while my heart was pounding. Then, like magnets, our mouths met.

Wacława broke free of the embrace and brought her finger to her lips, turning one ear toward the door. The Pole from my table, she whispered, had been pestering her; she suspected that he had followed us to her cabin and was now listening in the corridor, which formed a small cul-de-sac off of the main one. She didn't go check; she went to the bathroom to get ready for bed. When she returned, she was wearing the same lavender pajamas that Hania wore.

In the morning I descended to my cabin stunned with guilt. A psychiatrist would explain my infidelity as an act of revenge, after the emotional pain I'd been through, but it was purely about pleasure, not revenge, which of course can give enormous pleasure, albeit not of the sensual sort. But I was not, consciously or subconsciously, trying to hurt Hania; this was, for better or worse, all about me.

Naturally, I used our stumbling-out-of-the-gate marriage to try to justify my behavior to myself, just as I attributed it, in part, to the strange nowhereness, the seemingly exempt limbo, of a ship in the middle of the ocean. (The passenger liner was a gift to philanderers looking not just for opportunities but also excuses.) I cursed time again—why hadn't this happened on previous crossings, when I'd been single?—and desperately wished one could rearrange it in life as one did in stories. But nothing relieved the heavy weight on my conscience, not even the sorry pride I took in conducting an affair in Polish—particularly since it was with a Krakovian.

Yet I returned to her cabin again after dinner, during which the Pole had asked me, smiling, if I'd gotten a good night's sleep.

I rarely saw my cabinmates. One, I learned from an American student who had boarded in Tilbury, was a very interesting man, a survivor of Auschwitz. This added another layer to my guilt. I had the travel writer's conviction, even after two-and-a-half years and countless conversations, that I needed to talk to still more people; that the most fascinating characters—the ones with the most remarkable stories, the most exquisite insights—were still at large. Years later watching *Seinfeld*,

I would identify with Jerry when he made out with his girlfriend during *Schindler's List*.

The next-to-last evening carried a "Cowboy & Indian" theme. A group of us gathered in one of the bars, including a beautiful young woman whose long straight hair was belted by a thin brown headband. Marta was headed toward a reunion with her fiancé, an American who had studied medicine in Poland. One of the ship's officers arrived, handsome in his dress whites, and he and Marta exchanged knowing smiles. As they departed, his arm around her waist, she gave us a resigned, don't-judge-me look.

On the last day of the crossing, a long line formed in front of the duty-free shop. I came upon it with a feeling of exasperation, especially since the item of interest was vodka. Couldn't these people wait twenty-four hours and buy it on land, a bit more expensively but without any hassle? Were Poles now conditioned to standing in queues? Then someone explained to me that they were all buying the bottles to give to the crew, their waiters and stewards who weren't entitled to the discount.

Solidarity at sea.

CHAPTER 6

Philly Days

I SAT AT THE DESK IN MY BROTHER'S OLD BEDROOM, UNDER THE GAZE of Samuel Johnson. (I had cut his picture out of a magazine and taped it to the wall.) Books and papers covered the side bed, while a radio, tuned to WFLN—which came in only at night—joined books on the shelf of the other bed's headboard. A heavy, black L. C. Smith & Corona typewriter, purchased at the Salvation Army Thrift Store across the street from the *Trenton Times* building, decorated the pullout arm of the desk as I composed sentences on a yellow legal pad.

I was writing a book about the pilgrimage. The walk, I had decided, could serve as a microcosm of my entire time in Poland. I envisioned a colorful account rich in people, conversations, descriptions, observations; it would encompass religion and politics, past and present, and show how in Poland they are all entwined. Every morning, after breakfast with my mother, I climbed the stairs back to the bedroom to write. After dinner, I drove across the river to read Henryk Sienkiewicz in the Lafayette College library.

Sienkiewicz was not a writer I was drawn to, despite the fact that he had started out as a journalist and become something of a travel writer, publishing accounts of his journeys around the United States and Africa. But he was best known and rewarded—the Nobel Prize in 1905—for his sweeping historical novels. The only one that interested me was *The Deluge*, the second volume of *The Trilogy*, for it concerned the seventeenth-century Swedish invasion of Poland and the heroic—some believed miraculous—defense of the Jasna Góra monastery. But, before plunging into his hearty prose, I would go to Periodicals and read the *Spectator* wags.

My routine was interrupted by a swollen gland. I had not been feeling well—I'd recently developed a fever—though I put it down to delayed physical shock from the exertions, and paltry meals, of the pilgrimage combined with the emotional stress of separation and adultery. My diarrhea, I told myself, was the no-longer-dormant product of a cup of foul well water I'd gulped before smelling it. At lunchtime one day, watching a soap opera on TV, I sat up when one of the characters was told that his swollen gland might be Hodgkin's lymphoma.

"That's why I don't watch soap operas," my doctor said when I told him about the scene. He gave me some medicine and said it would soon go away.

It grew larger. I became reluctant to leave the house. When I did, I looked obsessively at people's necks, both sides beneath the jawline, searching for a bulge. No one else had one. I envied the men their unstretched collars.

I returned to the doctor, who this time wore a look of concern. He wanted to get a biopsy, he said, and scheduled me for surgery.

The mildly unreal sensation of being home suddenly became more so. It had been hard enough moving back in with my parents, finding my world suddenly emptied of friends, colleagues, students, Hania. On my first visit to the doctor, searching in my wallet for my insurance card, I had found a slip of paper with the new Boston address of Anna and Jan. It was a reminder that, a short time ago, I had had a full, engaged, eventful life. How could it have changed so dramatically and so fast?

Also, I'd often been sick in Poland—colds and flus and even bronchitis—but I had never, there or here, had anything that could be life-threatening.

I called Hania with the news—a rare, long-distance conversation—and was selfishly buoyed by her stricken tone.

On Saturday morning my father drove me to Warren Hospital. A blood sample was taken, and I was wheeled in for the outpatient procedure. The surgeon, a sprightly Filipino, called it off as soon as he saw the size of my gland. I'd have to come back, he said, and go under general anesthesia.

We drove home for a joyless lunch. I couldn't write. I couldn't even read, though I stretched on the sofa in the living room with a collection of stories by Bolesław Prus. In midafternoon the telephone rang; I got up and answered it in the kitchen. It was my doctor; the hospital had called him with the information that I had mononucleosis.

I ran to the den to tell my parents and then made another call to Hania, whose delight didn't erase her anger at my doctor for not testing me himself. I was a bit upset with him as well for being swayed by a patient who had self-diagnosed after watching a soap opera.

As often happens when you have a condition you've been assured is temporary, nothing changed. The swelling in my gland lingered, as did my diarrhea. (I had quickly gotten used to low American toilets again; Polish ones made you look like *The Thinker*.) Dan visited me during my convalescence, an act of friendship that touched me deeply. And a letter from Hania arrived like a balm. How beautiful those envelopes looked— with their artful jumble of chiaroscuro stamps, deep-blue airmail labels ("*POCZTA LOTNICZA*"), and unforsaken schoolgirl cursive—when plucked from the mailbox! Sometimes I didn't open them right away; I let them sit on my desk like rare finds, precious gifts, sealed mysteries, while I savored my anticipation of the sentences inside. This one didn't disappoint, as Hania had drawn in the body of the letter a cartoonish image of me lying in bed with a sickly expression. The title she gave this work carried the pet name she'd coined for me in Trenton prefaced by the noun-turned-adjective "Mononucleosis," and I saw it as the clearest sign yet that her affection had returned.

I resumed my regimen of writing and reading. Getting published in the *North American Review* had allowed me to call myself a writer; this—sitting at my desk every day, writing without deadlines or immediate hope of publication, watching with satisfaction as the pile of pages grew—made me *feel* like one.

When I finished *The Deluge*, I started Hilaire Belloc's *The Path to Rome*. There were not a lot of books about pilgrimages; this was still four years before Paulo Coelho's *The Pilgrimage* would initiate a torrent of them—but only if, as though by publishers' decree, they were about *El Camino de Santiago*. In any case, Belloc was much better company than Coelho.

My parents, while doubtful, were extremely supportive. My father wondered if I were considering converting to Catholicism, a move he assured me he would have nothing against. The question surprised me; my admiration for the Church, which had obviously come across in my letters, had more to do with its political than its pastoral role. One evening when I was still bedridden he brought me a copy of Neal Ascherson's *The Polish August*, which he had found on a rack in the drugstore. (Phillipsburg had no bookshop.) It says a lot about Poland's prestige at the time—which I took as encouraging—as well as the quality of titles that landed in unexpected places back then. It was like hearing of *Eothen* on a television talk show.

Like most fathers, he wanted to help his son, and he felt frustrated about having no contacts in the world of publishing. The closest he came was a slight familiarity with Richard Reeves. Reeves, then a political writer for the *New Yorker*, had cofounded the *Phillipsburg Free Press* after working at Ingersoll-Rand in the early '60s and deciding that engineering wasn't for him.

He returned to town to give a talk, and my father and I drove out to Harkers Hollow Country Club to hear him. On the way, I told of the talk I'd recently gone to at Lafayette, given by the novelist David Bradley. Bradley had started his career in a low-level position at a major publishing house in New York City, where his job had been to read all of the manuscripts sent in "over the transom" (unsolicited). He said that, because there were so many, it was impossible for him to read them from beginning to end. So he devised a method: He read the first page, the last page, and a random page in the middle. If none of them impressed him, he rejected the book. My father found his system as disheartening as I did, but, thinking about it some more, I saw the logic. Though it's often difficult to appreciate the final page if you haven't read all the ones that precede it.

In the country club's dining room we sat with two men my father knew. During the meal, he complained to them that much of what one read today was "too esoteric."

"I'm sorry, Howard," one of the men said humbly but also a bit cuttingly, "but I don't know what 'esoteric' means."

My father paused. I wanted to jump in—especially since this was a word he'd heard me throwing around—but I couldn't come up, quickly, with a good definition. More than most words, "esoteric" means what it is. I sat there frantically searching for something to say, to save my father, but words wouldn't come. It was the silence before *l'esprit d'escalier* that here had a vague air of patricide. Finally, my father said, "To tell you the truth, Charlie, I don't either." I stared down at my plate, unable to witness the humiliation I had caused.

The talk was entertaining, and afterward I approached Reeves, who was friendly and gracious. I told him about my time in Poland and my book about the pilgrimage, and he told me to write to him at the *New Yorker*.

Slowly, my health improved. I went by myself to see *Sophie's Choice*. I had carried the book around with me that summer in Princeton, getting curious looks from Jane, the French scholar, who took a dim view of bestsellers, as I normally did. But William Styron's novel had attracted me for obvious reasons—the story of a young writer infatuated with a beautiful Polish woman—and its more complicated and ultimately disturbing plotlines held me in their grip despite the verbosity and occasionally overwrought writing. Nabokov, I was sure, would have dismissed the book as *poshlust*.

The movie was an atmospheric, exquisitely acted adaptation, and, as the credits rolled, I stayed in my seat to watch it again. For two-and-a-half hours Meryl Streep had transported me back to my Warsaw classrooms, sounding remarkably like my female students—and I needed to hear them some more.

The first weekend in December, my brother Bill and his wife Pat took me to Philadelphia for the Army-Navy game. The night before it, we attended a party where I struggled to hold a simple conversation. It was with a woman who told me she worked in "software." Confused, I asked her to repeat the word. I knew about hardware but had never heard of its presumed opposite. I had been looking forward to functioning in English on a full-time basis again, and now I found myself unable to understand something as basic as a person's profession. In Poland, hardly anyone talked about work; asking people what they did for a living was considered almost rude. *Had* I inquired, it wouldn't have helped me now, as computers were nonexistent, at least among the general population. I had spent two years in a country waiting for a revolution while one of a very different sort had been taking place back home in my own. As someone who still wrote in longhand, I was not enthusiastic. A few weeks later, visiting Mike and Beth in their new home in Montclair, I told them, with the hollow confidence of the overwhelmed, that I was just going to ignore "this whole computer thing."

Walking to Veterans Stadium the next day, we came upon a small group of people protesting the war in El Salvador. They reminded me of Poland, that world of engagement, while the fans heading past them, oblivious or dismissive, told me I'd returned to the land of bread and circuses.

A political commentator once wrote, possibly in the *Spectator*, that the fact that a small percentage of Americans vote, rather than being a disgrace to the nation, was a sign of its health. It was in corrupt, tottering countries, he noted, that large numbers of people felt compelled, out of desperation, to take a stand. This certainly rang true with regard to Poland. But, as a superpower, the United States was different: It was a stable country that had the ability to create instability in others, a fact many of its citizens preferred not to recognize.

Though, a little weary after the last two years, a part of me welcomed the prospect of a meaningless spectacle, a large stadium—beyond the sections of cadets and midshipmen—filled to the brim with the politically carefree.

There were other annoyances to being home. Madonna was a big one. Not her music so much—which I avoided as best I could—but her name. For two years, "Madonna" was a term I had heard uttered with reverence. Madonna, the Blessed Virgin, was a symbol of suffering and sacrifice, pain and redemption. She was looked to as the succor of a nation and as the embodiment of its hope. She was the sacred protectress toward whom the pilgrimage I was writing about made its way. That a pop singer had had the temerity to usurp her name was bad enough; that she glorified sex and materialism seemed like blasphemy. Nothing illustrated the difference between the country I had left and the country I had returned to better than their celebrated, antithetical madonnas.

AIDS was another, more disturbing discovery. It was just starting its devastation of gay communities and causing panic in all sexually active populations. The timing of my mono, I noticed on reading news stories, seemed in tune with that of the fevers that people with HIV sometimes experienced. Of course, the possibility of my having contracted the disease from a female immigrant from Poland was highly unlikely. But it couldn't be discounted, at least not by me, with my weakened constitution and plummeting sense of self-worth. Guilt and hypochondria are a potent combination.

And then there were the rejections that were now trickling in. Another sign, albeit a less inflating one, that I was a writer. I had been sending proposals out to agents and publishers, who were now responding with a universal lack of interest. Most were form letters, or notes, written in the industry's generically polite and impersonal style—which I was now maddeningly familiar with—but some were specific. A few of these stated that there would not be a large audience for a book about a pilgrimage in Poland. The last two words, in retrospect, seemed to anticipate the coming onslaught of books about the one in Spain. But, even without knowing this, I sensed the innate unattractiveness of Poland—despite its two-year run on front pages—and got the impression that I was in for a long slog. The loneliness of the lover of the unsung. Even the pilgrimage I'd walked on was considered to be subpar. One afternoon I received a rejection whose unfeeling tone, after weeks of rebuffs, stung me so deeply that, carrying it up the stairs, I fell to my knees and pounded the steps as despairingly as I had pounded the walls that day in my bedroom in Arta. My romantic life and my professional life were both unhealthily dependent on mail.

Yet I made it back to my desk and, even though it seemed increasingly hopeless, I continued writing.

I wrote the last sentence one evening in February. I didn't usually work after dinner, but I was close to the end, a long passage of reported speech that I needed only to copy from my notebook. God bless the Dutch. When that was done, I scrawled "The End" with a flourish, something that's impossible to do on a keyboard.

I got in my Datsun—which in my absence had developed a crack in the dashboard—and drove to a bar for a silent, celebratory beer. I liked what I had produced, but I had the travel writer's eternal doubt: Is it really interesting, or does it just seem interesting because it happened to me?

Now every morning, instead of returning to my bedroom, I drove to my father's office on South Main Street and typed in the front room with his secretary. I never thought of asking her to do it; I didn't have the money to pay her, and she would have been incapable of reading my handwriting anyway, especially the insertions that cluttered the margins like lettered kites tied to inked-out lines. In the evening, sitting in the den, my mother proofread the day's finished pages.

After a few weeks, the typed manuscript was ready to be sent out. The neat pile of white pages looked more impressive than the pile of curled yellow ones did, and—in a way—more impressive than my copy of *New Jersey: Unexpected Pleasures*. This was my first real book—and the last to be written all of one piece, from beginning to end, until this one.

Immediately, I discovered the obsession that consumes people who've written a book on spec, without anyone requesting it. They appear to be normal, they're capable of engaging in intelligent conversations, but a part of their brain is always thinking about their unpublished manuscript. Its unformed state, its nebulous status, weighs on them constantly.

The Poles have a phrase for this phenomenon: *pisać do szuflady* (to write for the drawer). In Poland, the contributing factors were more political than qualitative, while here they're more economic. What is the difference between communism and capitalism? The former system bans books for their ideas while the latter bans them for their (perceived) inability to make money. This situation turns unpublished writers in the first system into dissidents and heroes and those in the second into poor schmucks.

Creators of manuscripts, the large armies of writers denied authorship—larger in the days before self-publishing—are incapable of forgetting the work they've brought into an unsuspecting world. The thought of it sitting unread dampens their happiest moments and turns their worst days even darker. Their lives will not be complete until their manuscript goes through editing (cosmetic surgery) and production (chrysalis) and emerges triumphantly as a book. And because of this, they never lose track of it—where it's been, where it is now—and they never stop searching for places to send it. They are constantly on the lookout for new agents, small publishers—someone, anyone, who might find them gifted and worthy of championing. In bookstores, they scan the spines of same-genre books and note the names of professionals thanked in the Acknowledgments. No potential lead is lost to them. Just as writers scour everyday life for material, unpublished authors comb it for markets. Grabbing the attention of someone known to the clubby world of publishing houses and literary agencies is always an option, but there were not a lot of these people in Phillipsburg.

Reeves wrote back to me, on *New Yorker* stationery, but, unsurprisingly, nothing ever came of his support. I wrote to Richard West, whose foreign dispatches I'd read in the *Spectator*, and he sent a letter saying that I "should and must" get the book published, an endorsement that lifted my spirits but meant very little to American publishers. My former Polish teacher introduced me to her friend in Princeton who was a friend of Styron—apparently, Joanna had been the source for much of the information about Poland that he had used in *Sophie's Choice*. She agreed to read the manuscript and then invited me to lunch, toasting my talent and my forthcoming success. That was the last I heard from her.

At a party in Princeton, I met a literary agent from New York, a young Englishman who said he'd be happy to take a look at my manuscript. I sent it off to him, which wasn't as easy as it sounds. I had to make another copy—my father had now exceeded his monthly budget for paper—and then place rubber bands to hold the pages together. After I found a box for it I put it inside, then wrapped the box, like a Christmas present, in plain brown paper. Carefully, I wrote the names and addresses of recipient and sender on the top. Then I drove it to the post office, where it was weighed and stamped and placed to the side while I searched for bills to cover the postage.

Several weeks later, the agent returned it—I assumed he had a secretary for the manual labor—with a note telling what the boomeranged manuscript showed: that he had declined to take it on. But he praised the writing, saying I had a good ear for dialogue, and suggested I try my hand at fiction.

It was exasperating advice. My so-called gift for dialogue was simply writing down, or remembering, what I heard people say. I may have been able to distinguish the amusing from the boring, the interesting from the mundane—as most people are; I wasn't capable of, or rather interested in, making quotes up.

I returned to the river road. Now, instead of heading into Trenton, I got on I-95, crossed the Delaware, and continued on into Philadelphia. Hania had decided to get her master's degree in city planning, and I thought I'd inquire about her chances at the University of Pennsylvania.

As at *Philadelphia* magazine, I arrived at the city and regional planning department without an appointment and simply asked if I could talk to someone. Seymour Mandelbaum promptly invited me into his office. He listened to my story with interest and said—as far as I recall—that they'd be happy to welcome Hania in the fall. She would need to send transcripts and such, but there seemed to be nothing to keep her from becoming a Quaker, joining in that club my father and brother. I continued chatting with the professor for another half hour. The City of Brotherly Love was also, apparently, the city of sympathetic editors and academics.

It seemed natural that I was going to live in Philadelphia, as it was where my beloved river led. I had always liked the city, not least because I owed my existence to it. Somewhere in its folds in 1941 my father, a student at Penn law school, had met my mother, a nurse at Children's Hospital. (Walking to work on inclement mornings, she would sometimes accept rides from C. Everett Koop.) As parents, they introduced my brothers and me to the zoo, the Franklin Institute, Connie Mack Stadium, Elfreth's Alley; as a student at Villanova, I took the Paoli Local into the city to watch Big Five basketball at the Palestra and, one memorable evening, strippers at the Trocadero Theater. In my junior year I bought my first pair of round tortoiseshell glasses—the same style I wear today—at Limeburner Opticians on Chestnut Street. Riding the train back to campus I'd imagine myself, in future years, getting off in Haverford, or Ardmore, or Bryn Mawr, or Rosemont after a long day of writing and editing in the city. I was not yet the urban creature I would become.

Philadelphia would be the first city since Aix that, as my place of residence, wouldn't be a fluke. Far from falling into it, I had been drifting toward it since birth. But it conformed, especially in the 1980s, to my ideal of the uncelebrated. I had never considered living in New York. New York was a colossus, a global curiosity, and, increasingly, a broken record of rejection; it was not a place that one called home. It astounded me that eight million people did. It was too impersonal—even its big river, the Hudson, intimidated, especially compared to the inner-tubed Delaware (at least in its upper reaches)—and far too self-important. As a travel writer, or at least a student of travel writing, I knew that the best in their fields did not always end up in Manhattan—Theroux at the time was living in England—just as I knew that not everything the city produced was superior. *Travel + Leisure* was a repository of pretty pictures and bland articles put out for consumers of luxury travel, not connoisseurs of fine writing. Owned by American Express, the magazine bold-faced the names of hotels and restaurants in its stories. Similarly pedestrian writing appeared in the *New York Times* Travel section, with the exception of its end-page essays, while the great *Holiday* had been assembled in a building on Independence Square.

Hania arrived in April, and I drove to Kennedy to pick her up. Before entering the Pan Am terminal, I tore into small pieces and deposited into a trash can the two perfumed letters I'd received from Wacława. In Phillipsburg, as we came to a stop in the driveway, my wife jumped out of the car and ran to the porch to embrace my mother.

My parents threw a party in celebration of Hania's arrival, after which, upstairs in the bathroom, I shaved off my mustache. Because I had grown it in France, eight years earlier, its removal seemed to mark the end of my traveling days, at least the years of deep, indelible, long-term travel.

When asked to give advice to young people looking to become travel writers, I invariably tell them to go—alone—and live in a country where they don't speak the language. They'll learn not only the language but also the customs, the songs, the jokes, the history. I point out to them that this is precisely what I did.

Well, yes, but for a different reason. In writing this book it has occurred to me that the majority of my foreign sojourns—my working abroad, my living with the people; what I now consider the ideal travel-writer apprenticeships—were undertaken not for my career but for love. My professional life was the direct beneficiary of my romantic life.

I also moved my wedding ring to my left hand but, because that's the one I use the most, I found it chafed and soon moved it back.

We spent the spring at Jim's summer house on Long Beach Island. It was the opposite of our idyll in New Hope. Despite her generous act upon arrival, Hania was not enthusiastic about finding herself in the States again, this time permanently, and she was anxious about our situation: no job for me, no apartment for us—both of which I was supposed to have found by now. I told her I had been writing a book about the pilgrimage, but there was no book to show her, just a stack of unwanted pages. The special melancholy of the seaside-out-of-season further flattened our moods.

We took day trips into Philadelphia, driving through the Pine Barrens, Early McPhee Country. The neighborhood that appealed to me was Queen Village because it had a view of the Delaware and its superannuated fleet, which included the *Moshulu*, the four-masted barque made famous by Eric Newby in *The Last Grain Race*. In fact, it sat between two suspension bridges as expansive as their namesakes: the Ben Franklin and the Walt Whitman. The only downside was that a highway—I-95—cut residents off from the river, producing not just a physical barrier but also a noisy one. American cities were late to discover the aesthetic value of riverfronts; in many, like Philadelphia, irreparable damage had already been done.

At one time Queen Village had been a Polish neighborhood—St. Stanislaus Church still stood on Fitzwater Street, Rachubinski Funeral Home on Front—a history that delighted me more than it did Hania.

While significantly cheaper than Society Hill to the north, it had many of its features: tree-lined sidewalks, redbrick row houses, streets that in newer cities would have been alleys. After Warsaw, Philadelphia—at least this part of it—resembled a miniature village, a rebuke to the idea that everything in America was bigger. We had moved from the monumental to the homey, and from the slipshod to the tidy.

We found a studio apartment on Catharine Street, between Front and Second, and moved in on a warm June evening. I walked a few blocks to call on a woman I'd worked with in Trenton—predictably, we needed to borrow something—and, on returning, I found that Hania had set up our camp bed and our single lamp in such a way that the place, or at least one corner of it, actually looked cozy.

The studio made our Warsaw apartment seem luxurious. All we had was a room—with exposed brick walls and two large windows—a bathroom, and a kitchen so narrow the door of the refrigerator couldn't open all the way.

A traveler for much of my adult life, I had rented an apartment based on the same criterion I had always used for choosing hotels: location. Forgetting, apparently, that one eventually checks out of a hotel. For years

my ideal had been to live within walking distance of a bar, a bookstore, and a bakery. A bar was visible from our windows, on the corner of Second and Catharine. South Street, four blocks to the north, was home to Book Trader, a spacious, two-story, secondhand bookstore. South Street also had a gourmet market, which sold excellent bread, and a diner, where I started playing chess in the evening with an old friend from Villanova.

I felt I was back with my people. I'd look at them on the street and see no mysteries: I knew where they had gone to school (I had gone to school with them), what they ate for lunch (hoagies and cheesesteaks), where they went at "the shore," which cities' teams they hated the most passionately. It was a pleasant feeling after four years of living abroad, though it wasn't the same as a sense of belonging. I was still a newcomer in a place where most families had lived for generations. This made it different from New York, or at least Manhattan, where settling in was easier, I suspected, because of all the other outsiders.

Hania found the city provincial compared to Warsaw, something few people understood. Wasn't Warsaw behind the Iron Curtain? Yet most American cities, with the exception of New York—and possibly San Francisco—were not what a European would call cosmopolitan. A store on South Street specialized in magazines—its racks crammed with weeklies, monthlies, quarterlies, literary journals, fashion magazines, gossip magazines—all of them in English, published in the United States. You could not find the *Spectator*, let alone *Der Spiegel*. I wrote an op-ed piece noting how the city was filling with ethnic foods while its newsstands uniformly excluded the foreign. I declared that Philadelphians' palates were far more sophisticated than their minds. This was years before the city became known for restaurants. I sent it to Phil—who had moved from Kentucky to the *Los Angeles Times*—and, after he published it, it got picked up and commented on by a columnist at the *Inquirer*.

A frequent winner of Pulitzers, the *Inquirer* had a high opinion of itself. When I had gone there to see about work, I had been told that I needed at least five years of newspaper experience to even be considered for a job. The fact that I had spent the last two years living, and writing, in a country going through political and social upheaval made no impression on the editors. They seemed to view me—despite my clips and my languages—as someone just back from a lark. Didn't Americans go to Europe for *vacation*? Mine had simply been a ridiculously extended one. And besides, foreign experience is much less important than career continuity—at least to a provincial.

I did get a freelance assignment from the paper. It was to write about the Polish community in Camden for one of the local, pullout sections. The neighborhood, like much of the city, was rundown, but I found a Polish meat market—*kabanosy* again!—and, of course, a church. The resulting story, which I was quite pleased with, bore little resemblance to the one that appeared in the paper. It was my first experience with aggressive editing, and it came as a shock. I had written for three people in my career (other than Phil)—a newspaper features editor, a magazine-turned-book editor, and the editor of a literary quarterly—and none of them had found my writing in need of much improvement. Now, suddenly, it was so unsuitable that whole paragraphs had had to be rewritten. On my L. C. Smith & Corona, I typed an indignant note to my editor, knowing it would kill any future assignments, but my professional pride took precedence. She wrote back saying that, if I was going to be a freelancer, I had better get used to editors who edit. As a prophecy—at least the bit about intrusive editors, not my learning to tolerate them—it would prove depressingly accurate.

Summer arrived, giving Hania her first real taste of heat and humidity. She found it as unpleasant as I had found Poland's cold and damp. The days here were brighter, but the skies were rarely blue; instead of one low-hanging cloud there was, higher up, a gauze-like haze. "Roll out those lazy, hazy, crazy days of summer," I hummed to myself. Nobody ever sang about overcast skies. Okay, Jacques Brel had, but he wasn't Polish.

My Datsun had no air-conditioning. I would drive with the windows down, trying to catch a breeze, and arrive at job interviews in pants wet with sweat. After a few days, they developed white salt marks in the back. Coming home, I'd pass fat cockroaches loitering in the passageway leading to our studio. They helped restore America's Brobdingnagian reputation.

Shopping for clothes, Hania was beguiled by the sheer quantity of goods—after the depleted stores in Warsaw—before discovering that it didn't translate into limitless choice. There was a frustrating lack of variety, as whatever was in fashion at the moment conclusively pushed out whatever wasn't. This phenomenon in the clothing industry carried over into the world of publishing.

In a store in Head House Square we bought a couch that doubled as a bed in the most rudimentary way possible: The cushion simply spilled out onto the floor. We now slept at a lower level than we had in our camp bed. On Sunday mornings I'd walk up to Society Hill and sit in a box pew in St. Peter's Church, where the priest gave his sermon from a wineglass pulpit. After the service, coffee was served in the parish hall on Pine Street, a few doors down from the house in which Kosciuszko had lived.

The church encouraged parishioners to get together socially and had established a program of home evenings. We hosted one for a necessarily small group that included a young social worker. Arriving in Philadelphia, Carol had found a small row house in South Philly and had immediately become an object of curiosity. Hardly anyone in the Italian neighborhood lived alone; frequently several generations shared one house. People felt sorry for her, being without family, and they became extremely protective. It was, she thought, the safest place she could live in the city. Whenever a man came to pick her up for a date, the local toughs would give him a thorough grilling before he got to her doorstep.

She grew close to her neighbors, especially the women, who often made use of her professional services. On Sunday mornings, when the men went to their meetings, their wives would come to her house and tell her their stories. One morning a woman was in tears, convinced that her husband had taken a mistress.

"Why do you think that?" Carol asked her.

"Because," she said, "he no longer beats me."

At Penn, Hania met a Polish couple who invited us to a meeting of the local Solidarity chapter. Elżbieta Sachs was a short, handsome, stylish woman who spoke, in addition to Polish and English, French, Spanish, and Swedish. She worked, appropriately, with the university's studies abroad program. Her husband Włodek, tall and droll, did something abstruse at Wharton. One weekend they hosted a party at their home in Wynnewood, where a young woman with round wire-rim glasses and thick brown hair sat in the sun like a self-possessed schoolgirl. She introduced herself, in a dulcet voice, as Agnieszka.

The Solidarity chapter gave assistance to new Polish immigrants, a few of whom—like Robert Terentiew and Krzysztof Hariasz—had recently been interned and enjoyed something of a celebrity status. Robert basked in it while Krzyś shrank from it. I started giving evening English lessons in Richmond, the still-thriving Polish neighborhood. The instant gratification of teaching—seeing a face suddenly light up with understanding—was the perfect antidote to the delayed rejection of writing.

At the beginning of October it was announced that Lech Wałęsa would be the recipient of the 1983 Nobel Peace Prize. Three days later my letter to the editor appeared in the *Inquirer*:

"Here are seven reasons why I agree with the *Inquirer's* decision to devote four columns of its front page to the Philadelphia Phillies' Oct. 5 loss to the Los Angeles Dodgers, and two columns to the awarding of the Nobel Peace Prize to Lech Wałęsa.

"The playoffs to the World Series are, as the title suggests, of a universal scale, including all nations of our earth in their influence, while the Nobel Peace Prize is not much more than personal approbation from a few anonymous men in Norway."

Reason #6 was: "World peace may be, indeed, a noble striving, but baseball combines, as no other human endeavor, the dual dynamics of hitting and pitching."

I was still seeing things with a foreign mindset. When Hania and Elżbieta and Agnieszka complained about Americans, which they did occasionally, I'd put up a half-hearted defense, only to be told, "You're not American." I didn't protest; I knew what they meant: I wasn't the kind of American—typical, in their eyes—who watched sports and drank beer and knew nothing about the world outside the United States. (Actually, I did watch sports and drink beer, though usually not simultaneously.) And their assessment struck me as flattering, for it seemed to place me in the cosmopolitan ranks of Europeans, where I wanted to be. Though I wondered if this desire, as it manifested itself in my writing, was holding me back professionally. I had long been attracted to, which meant I was heavily influenced by, British writers, not just in the field of travel, where they excelled, but in the realm of succinct, subtle, dryly humorous prose. And this put me at odds with the American penchant for rambling, word-drunk, often overly earnest texts. The British tendency was to hold things back while the American one—beginning long before the '60s—was to let it all out. I much preferred the Latinate sentences of Waugh to the overstuffed ones of Wolfe—Thomas and now Tom. Of course, Fitzgerald had written beautifully measured lines, and Hemingway's had had a revolutionary leanness, but our contemporary writers—from Mailer to Styron to Bellow to Irving—were all enamored of the sound of their own typing. (As exquisite as, in Bellow's case, it often was.) In travel, Chatwin had a lapidary crispness that Theroux, for all his Anglophilia, lacked. Instead of understatement, the Americans gave me gonzo.

Yes, the editors at the *Inquirer* published my letter, but they didn't offer me a job out of admiration for it. The sardonic voice that had wowed them in Trenton, of all places, failed in Philly.

I continued reading the want ads and, one day, saw that the American College of Physicians was looking for a writer. If journalists wouldn't have me, perhaps doctors would.

The ACP, I learned during the interview, put out an in-house magazine called the *Observer*. The editor, Bob Spanier, was about my age and a fellow Jerseyan; he was smart and funny and intrigued, I suppose, by my work history. Personal chemistry is as important in getting a job, especially in small organizations, as background and expertise. In this case, it was more so.

The ACP was then headquartered in an expanded mansion in West Philadelphia. The *Observer* provided members with a monthly supply of news and feature stories, and writing for it was a job that had the potential to teach me a lot while getting me nowhere. The only people who read the *Observer* were doctors of internal medicine, and they didn't read it because they were too busy keeping up with the medical literature. I found this more disheartening as a staff writer than as a future patient.

Fall is a good time to start a new job, at least in the States, as Thanksgiving is quickly followed by Christmas. The ACP office party was held in a hotel just off the Penn campus. When "Ave Maria" played, one of the women who worked in the finance office asked one of the doctors who was on the board of regents to name the composer. "Schubert," he said, without hesitation. A few minutes later, Frank Sinatra came on, singing "New York, New York," and the young people got up and sang along with him. I stayed seated. The song was in close competition with "My Way" as my least favorite in the Sinatra repertoire—most of which I now loved. But more depressing than the music was the sight of Philadelphians exuberantly singing a hymn to the city in whose shadow they dwelled. It seemed to reveal an inferiority complex that—unlike most—was happily acknowledged, and shamelessly demonstrated. I would have preferred a group singing of "South Street" ("Where do all the hippies meet?").

After the party a few of us went to dinner at an elegant Chinese restaurant on the street the hippies had long abandoned. Sitting there with my new colleagues from my new job not far from my new home—the energy of the city mixing with the ebullience of the season—I delighted at being part of something again. Especially after my solitary winter of writing. That evening, along with financial considerations of course, would be partly responsible for my always preferring a job to a freelance career.

New Year's night, looking out the windows of our apartment, I saw two silvery mummers staggering down Second Street as if through a dream.

In February, we moved to a one-bedroom apartment at 47th and Larchwood in West Philadelphia, gaining not only space but also proximity. Now I walked to work—an American novelty—and Hania took the bus to Penn. Through the wood floor we sometimes heard our downstairs neighbor playing the piano; Leslie was a student at the Curtis Institute of Music and her cat, Mieczyslaw, was named for her favorite professor.

I returned to our old neighborhood frequently, often to scour the shelves at the Book Trader. On the second floor—travel sections are never centrally located—I found the works of George Borrow, Gerald

Brenan, Peter Fleming, Laurie Lee, A. J. Liebling, Jan Morris. I discovered more works by Freya Stark—one of which, *Beyond Euphrates*,
carried a replica of the author's signature in gold on the Philadelphia
Eagles-green cloth—and landed a beautiful, faded, dust-jacketed copy of
Isak Dinesen's *Out of Africa*. Kate Simon was well represented with her
"uncommon guidebooks," which I also scooped up. They mixed practical
tips with lyrical essays and had relieved travelers in the '70s—at least the
ones headed to Mexico, London, Paris, and Rome—of the necessity of
packing any other books, for, in them, helpful information and pleasurable reading sat side by side. Now her write-ups of hotels and restaurants
were often outdated, but the prose was fresher and more illuminating
than what one found in the latest issue of *Travel + Leisure*. Introduced
by Fussell to Douglas and Byron—whose name I had first come across
in Waugh's introduction to *When the Going Was Good*—I now added *Old
Calabria* and *The Road to Oxiana* to my growing travel library. Here and
elsewhere I acquired more classics—*The Innocents Abroad, Travels with*

a Donkey in the Cevennes, Down and Out in Paris and London—as well as works by other novelists-cum-travel writers: George Gissing's *By the Ionian Sea*, Arnold Bennett's *Those United States*, D. H. Lawrence's *Mornings in Mexico*, Aldous Huxley's *Along the Road*, Graham Greene's *The Lawless Roads*. On my shelves at home I put Stendhal's *Travels in the South of France* next to Goethe's *Italian Journey*; Hilaire Belloc's *Places* next to Henry James's *Portraits of Places*; Karel Čapek's *Letters from England* beside George Mikes's *Little Cabbages*, in which he did to France what he'd already done to Britain. The writers were more important to me than the places they wrote about; geographically, I had no real biases apart from an impatience with tales of exploration. In books, as in life, I needed people, craved civilization. But the whole point of the exercise was to learn how to write; learning about the world was a concomitant benefit. And one that didn't always come from reading fiction.

I soon tried the *Inquirer* again. It was a large newspaper, and I hoped that communication between departments was not unlike that between the visa office in Warsaw and the embassy in Athens. In my spare time I had written a story about my Easter in Greece, complete with a description of the bus ride to the island of Trizonia, and I confidently sent it to the travel editor. A few weeks later a letter arrived in my mailbox:

"Dear Tom," it began. "Someday when you write your first successful book you will remember me and chuckle and say, 'Take that, asshole.' Meanwhile . . ."

After years of "We appreciate your interest, but unfortunately your story is not right for us at this time," this stood out. It perfectly captured my thought process, minus the obscenity. But it was also confusing. It suggested I had the talent to write a book—but not a newspaper article, at least not one for the *Inquirer*'s Travel section. What kind of talent was that? And the fact that I *had* written my first book, without success, had the unintended effect of making the flatterer sound as if he were rubbing it in.

The letter went on to state the editor's main objection to the piece, which was that it lacked a theme. "Is the story about Greek Easter, the little island, or getting to the little island?" he asked. He had a point, which I probably didn't acknowledge at the time, so convinced was I that stories should be judged on their writing alone. Years later, as a travel editor, I would include the absence of a theme as one of my "10 Sins of Travel Writing"—which, for a while, I sent out with rejection letters—while at the same time publishing stories based solely on the strength of the writing.

One day I sat down and wrote a letter to V. S. Pritchett, which I then sent to his publisher. In it I told him how, arriving in London in the fall of 1976, I had gone straight to Foyles and purchased *Foreign Faces*, then given it to the Polish barmaid at my hotel, a gesture that led to my living for two-and-a-half years in Warsaw and for five years and counting with the barmaid as my wife.

A few weeks later a pictureless postcard appeared in our mailbox. On the back, at the top, were printed the words: "From Lady Pritchett, 12 Regents Park Terrace, London, N.W. 1." The word "Lady" had been crossed out and the initials "VS" written above it. The message below flowed in shaky handwriting: "I am so glad to hear that a book of mine has been enjoyed by you and with such delightful after-effects. VSP." After showing it to Hania, I placed it in the talismanic volume.

I also started work on a new book about Poland. It had become clear to me that the pilgrimage was doomed to the drawer, so I embarked on a longer story about my two-and-a-half years in the country. I wrote it in separate parts that could be stand-alone essays—one on Warsaw, one on the English Language College, and one on martial law, which, I had decided, would consist, in part, of excerpts from my journal. The longest chapter would be about the pilgrimage, after I went through the painful process of cutting what I'd already written by about two-thirds—killing not just my darlings but hundreds of pages. My goal was to create a book like the one I had searched for before going to Poland, one that told about the people, their culture and customs and everyday life. It would satisfy a need, I thought, and stand out from the crowd of political books that were now appearing.

On Good Friday I extracted from our mailbox a large envelope that carried the return address of *Commonweal*. I had sent the Catholic biweekly an excerpt from my book, about the first Easter during martial law, and I assumed the editors were returning it, in annoyingly timely fashion. Opening the envelope, I pulled out that week's issue, with a list of articles on the cover. Among them was "Warsaw Easter" by Thomas R. Swick. Eagerly turning inside, I found my story, graced by caricatures of Wałęsa and Jaruzelski. It ended with my description of listening with Hania's family, on Easter Monday, to Solidarity's first underground radio broadcast—a broadcast that had had parallels, for me at least, to the story of the Resurrection.

The boost to my confidence didn't last long. I still had no leads, no interested parties for the book that story was a small part of, and I fought through doubt and an almost constant feeling of futility. Sometimes, instead of writing, I'd curl atop the bed—our bedroom was my office—in a sad pile of listless self-pity. Hania showed no sympathy; on the contrary, she'd come in and her face would register disapproval, tinged with disappointment. It's hard to convince a Pole, even a loved one, of your bleak circumstances. And even if they are bleak, you need to show your resolve and overcome them.

Happily, I didn't mind going to my dead-end job. A new assistant editor had joined the team, and everyone got along extremely well. The editorial assistant, a young woman from Reading who played in a punk-rock band, had begun reading the *Spectator* magazines I brought into the office and had taken a liking to Jeffrey Bernard. And I enjoyed the work when I wasn't covering right-to-die issues, which seemed to come up about every other month. At Penn I interviewed Renée C. Fox, the brilliant medical sociologist and bioethicist who wrote a great deal about organ transplants, especially of the heart. Because of that organ's symbolic importance, we heard more about heart transplants than liver transplants, which, Dr. Fox told me, were much more difficult and complex operations. She was the first person I heard use the word "leitmotif" in conversation, and it took me a while to match her pronunciation with the inscrutable letters I'd seen on the page. A woman at the university's Museum of Archeology and Anthropology stopped in the middle of a sentence and said, "I apologize if I'm just burbling on"—a phrase I hoped to get into my story. I took a train to New York City to interview Victor Herbert, a doctor named for the famous composer (his father's cousin), who, in his fight against quackery, had obtained a medical degree for his cat. I returned to the city to write a profile of Calvin H. Plimpton, an ACP member who was on his way to Lebanon to serve as president of the American University of Beirut. He was a tall, jowly, likeable man who, unlike many of his colleagues, didn't take himself too seriously. When I asked him how he had chosen medicine, he said that he had suffered from asthma as a child. "When you spend your childhood having other people stick needles in you, you begin to think it would be fun to attack." I took some pictures of him to accompany the story, and, when they didn't come out, I drew a caricature. The executive vice president, a normally laid-back doctor who brought his German shepherd to work, was not pleased when the issue appeared, but I never heard a word from Dr. Plimpton. Lebanon surely gave him more important things to concern himself with, and he struck me as someone who wouldn't mind being caricatured, even by an amateur. Once I took the train to New York City and then the Hampton Jitney to Amagansett, where I interviewed Berton Roueché, writer of the *New Yorker*'s popular "The Annals of Medicine." We sat and talked in the house where, several years later, he would take a shotgun and shoot himself in the head.

In many ways, it was like writing features back in Trenton, but in a circumscribed field and without the community of aspiring writers. Now, instead of the *New Yorker*—or, actually, in addition to it—my boss read the *Morbidity and Mortality Weekly Report*.

The *MMWR* was published by the Centers for Disease Control, where I was sent one year to cover a conference on preventive medicine. It was my first trip anywhere on someone else's dime. Looking at the distant skyline of Atlanta from my hotel room, I felt as if I'd entered the travel writers' wading pool. I wrote nothing about the city, of course, though I did interview Jimmy Carter. I regretted not bringing the birthday card I had drawn in Arta.

I returned to Atlanta the following year to write about a professor at Emory who had his medical students read literary classics. Again I boarded a plane without paying for it, which seemed like a small miracle, and found the class discussing *Madame Bovary* at Manuel's Tavern. In preparation, I had reread the novel I'd studied in college—in a French literature in translation course—and marveled anew at the writing, sentences that seemed never to carry an unnecessary word.

There were annual trips to the ACP national meetings. The first I attended was in San Francisco, a city I had wanted to visit since childhood when, as the family rebel, I had been a Giants instead of a Phillies fan. Like the doctors, Bob and I were put up in the Fairmont Hotel, where, one afternoon, I interviewed Lewis Thomas, the physician essayist who had won a National Book Award for *The Lives of a Cell*. Our conversation was more about writing than it was about biology. During the day I explored the city, and at night I listened to the pianist in the hotel bar.

"Splat," she sang, in a song that paid homage to the *New York Post*, "a man fell fifty stories on his back. Flat." I went up to her at the break and expressed my admiration for her songwriting talents. We talked a bit; I told her of my reason for being in the city, and she said she had been asked to play at the ACP's closing dinner. I invited her for a drink afterward.

During the dinner she played mostly standards. She was tall, with curly black hair and a long, thin, freckled nose. When the evening came to a close, I walked up to the piano and helped her on with her jacket—astonishing, I'm sure, the assembled physicians, at least the ones who knew me. We marched out of the ballroom, through the lobby, and across the street to the Mark Hopkins Hotel. She was a fan of its penthouse bar.

Over cocktails, I listened to her stories of metropolitan life while occasionally gazing down at the Kasbah-ed city. At one point she asked about my hybrid accent; I told her of my years teaching English abroad. "It's like having drinks with Cary Grant," she said. It made me wonder if I could now use my voice to hide my awkwardness. (We were still a decade away from the floppy-haired bumblings of the other Grant.) We left around midnight and said our goodbyes down on California Street. Returning to the Fairmont, I felt a sense of accomplishment; I had not only seen San Francisco but met one of its inhabitants. I had had an experience of the city that the average tourist doesn't have—the mark, it seemed to me, of a good travel writer. Even one who isn't writing a story.

Two trips I took for the ACP had even more of a travel aspect. One was to Chicago to write a story about Cook County Hospital. It was the archetypal big-city hospital and, while I didn't like hospitals, I loved big cities. Every day I took the train from downtown to West Harrison Street, where I joined the homeless, the indigent, the teenage mothers, the wearers of casts and the hobblers on crutches as they maneuvered their way through the nonautomatic doors of the fading Beaux Arts building. Inside, I interviewed doctors about the legendary teaching hospital. It was here, I learned, that the world's first blood bank and the country's first trauma unit were started. Contrary to what a BBC documentary had reported, there had never been gang wars in the tunnels. "We used to have big rats in the tunnels," one of the doctors told me, "but never gang wars."

Back in Philadelphia, I gave the story an atmospheric lead—describing the demonstrators and the man with the *helados* cart outside the entrance—and, halfway through, inserted a passage from Liebling's "Chicago: The Second City." It ran to more than three thousand words and was read, quite possibly, by some of the people who were interviewed in it.

My most enjoyable trip was to San Angelo, Texas, to write about a doctor who had recently returned from a mission providing medical assistance in Afghanistan. This was the mid-'80s, when the mujahideen were fighting the invading army of the Soviet Union, a good while before they would turn against us.

Why I didn't just do a phone interview I have no idea, but I was delighted that our budget allowed for these extravagances. Before leaving I called Dr. J. Preston Darby, who offered to come pick me up in Midland. I asked how far that was from San Angelo. "About a two-hour drive," he said, before assuring me that that was nothing in Texas. I told him I'd take a small plane into San Angelo.

He drove to my motel, and we did the interview in my room. (I hadn't rented a car.) He fit my preconceived image of a Texan, especially one who'd been smuggled into Afghanistan: tall, rugged, self-assured. He was also eloquent and gave me good descriptions and details that would help make my story as vivid as a secondhand account could be. And his patriotism—or was it regionalism?—came through in his monologues.

"When I was tired and couldn't keep up with the Afghans, they would say things about Americans being soft. Americans for them are people who have a very easy life. One day I got mad about this and I said: 'Yes, that's right. We have it easy. But my grandfather didn't have it easy and my great-grandfather didn't have it easy. But they didn't sit on their asses around a lake and not build boats. They worked for some improvement in their lives. And you people, if there weren't a war going on, should get up off your fannies and stop defecating in your fields and build privies and have a better life and enjoy it. I haven't seen a privy in a single village. You want me to treat these kids for worms and if I do they're gonna re-infect themselves in a few days. Tell these people to build a privy. Tell them to wash themselves, to take a bath every day. You start doing this and your country will progress. To live in your own misery for centuries is stupid.'"

When he had finished telling his story, Dr. Darby took me for a ride. I had called the motel receptionist shortly after checking in and inquired if there was a main street I could walk to. She had mentioned the name of a chain restaurant. "But is there a downtown?" I had asked her. "Not," she had said slowly, "as you and I would know it."

We passed groupings of mesquite trees—grilling with mesquite chips had become popular back East—and a small convention center that was hosting a show of barbed wire. The doctor spoke fondly of his fellow West Texans; they were, he said, stalwart people and true to their word. But the routine of a country doctor got to him sometimes. "Was it Thoreau," he asked, "who said most men live lives of quiet desperation? I spend my days sticking my finger up men's asses." I sensed that, for all its Good Samaritanism, the trip to Afghanistan had also answered a need for adventure. The doctor had used travel not as an apprenticeship, or a path to romance, but the way most people did: as an exhilarating break from routine.

We dropped by his house, where I met his wife. Still taking pity on me, he invited me to join them for dinner and a movie. It was, after all, Saturday night. Before the picture began, a message appeared on the screen: "PLEASE REMOVE TALL HATS."

I returned to Philadelphia with two stories to tell; the Texas one, in fine Western tradition, I related orally.

In the fall, Hania and I made our first return to Poland. I got very emotional as the plane descended and the clouds parted, revealing heartbreaking hives of familiar gray apartment blocks. It felt strange walking through the arrivals hall and not looking for Hania, having her instead by my side.

Jaś and Elżbieta were doing okay, as well as could be expected. Jaś had replaced his Syrena with a Ford, which he'd bought in West Germany with the petrodollars he had earned in Iraq. The mood of the country was subdued; after the defeat of Solidarity, Poles who hadn't immigrated had retreated into themselves. A number of our friends had started small businesses. This worked doubly to the government's advantage; the entrepreneurs were often giving society things that the state had failed to provide, and they were doing it individually, or in small concerns, so there was less threat of organized political opposition. One woman told me she sometimes felt nostalgic for the old food lines. "At least there we were as one," she explained, "bitching and complaining together. Now we're all dispersed, and we go about the shops without a word."

I missed the hopeful energy of the Solidarity period but, focusing my attention on the current situation, which I planned to write about—working vacations are all the budding travel writer knows—got me reflecting on its applicability to my own life. Like the Poles, I needed to persevere through a difficult time.

Back home I wrote an essay about the trip. It was not really a travel story, more a work of reportage. It was long and leisurely, with detailed descriptions of Warsaw and meetings with Poles, some of which included quotes (see above). Its most commendable aspect, perhaps, was that it carried the voice of someone with a deep familiarity with the place.

I gave it the title "Private Lives," reflecting the theme I had dutifully found for it, and sent it to the usual places—the *New Yorker*, the *Atlantic Monthly*, *Harper's*—as well as to the *American Scholar*.

I had been introduced to the *Scholar* by Vernon Young, a freelance writer whose work occasionally appeared in its pages. Vernon had come to Philadelphia to escape the high rents of New York at the suggestion of an editor at the *New York Times Book Review*, a woman by the name of Rebecca Sinkler, who commuted to Manhattan from her home in Center City. I had read his review in the *Inquirer* of Theroux's *The Kingdom by the Sea* and had sent him a letter, care of the newspaper. When he responded, I invited him for tea.

I met him on the stairs, where he had paused to catch his breath. His tall, thin frame seemed ready to collapse under the weight of his herringbone overcoat. Watery blue eyes topped a veined Roman nose, and wisps of gray hair sailed back from a high forehead. A plummy accent reinforced my impression that he had stepped out of the pages of *New Grub Street*.

Vernon was a critic, primarily of film. His reviews appeared regularly in the *Hudson Review*, and he was the author of *Cinema Borealis: Ingmar Bergman and the Swedish Ethos*. But he also wrote about literature and was as comfortable pronouncing on poetry as he was on travel writing. His broad knowledge and deep intellect impressed me, and Hania, though she found his precarious life, living from review to review, depressing. While to me he was an old-fashioned man of letters, an almost romantic figure—he had even written a novel, *Spider in the Cup*—to her he was a cautionary tale. I couldn't even hint about quitting my job—which I did sometimes during grounded spells—without Hania asking, "Do you want to end up like Vernon?"

But I coveted the life that had gotten him to this unenviable state. He had known Federico Fellini in Rome, before he became famous, after which, Vernon said, you could no longer have a normal conversation with the man. There was something about fame, and riches, and sycophants that made him emotionally unreachable. Researching his Bergman book, Vernon had lived in Sweden, an experience he had still not recovered from. You'd be invited to someone's house for dinner, he said, the snow accumulating outside, and, when the evening was over, instead of offering to drive you to the train station, they'd blithely wave goodbye to you from the warmth of their vestibule. "In every country one has to learn to live without something," Vernon proclaimed, in one of his frequent aphorisms. "In Sweden, it's people."

He reminded me of a more bookish Jeffrey Bernard, the Englishman stylishly going to seed. He didn't drink, at least not to get drunk, but he smoked, and the scent of cigarettes permeated his clothes, his hair, his whole essence. Like a lot of once-dashing ladies' men, he continued to see himself in that role. Once he brought to a movie a blond graduate student who had contacted him about a paper she was writing. About a month later, I asked him how she was doing. He looked off into the distance for a full five seconds, considering the question. Then he said gravely: "I'm still in the picture."

He lived on Spruce Street in a one-room apartment I could only imagine. He had a few writer friends in town, including Jerre Mangione, who wrote about the Italian American experience and had been active in the '30s in the Federal Writers' Project. He was in much better financial straits than Vernon, and I wondered if this was due in part to the fact that he had a wife. We'd all meet sometimes for simple dinners out; once, I watched Jerre finish off a plate of spaghetti without touching his bread. When I asked him why, he said, "It would be redundant." I have not eaten bread with pasta since that night. And I still have my copy of *The World Around Danilo Dolci: A Passion for Sicilians*, with the inscription, "To Tom Swick—With a toast to his travel writing future."

It is dated August 10, 1986. Two months earlier, my essay on Poland had been published in the summer issue of the *American Scholar*. Not only were my words inside it, but my name appeared on the pale green cover.

I had been sending the *Scholar* excerpts from my Polish book, which the editor, Joseph Epstein, had turned down thoughtfully and in tones of encouragement. From his comments I took away some hope of eventual success.

News of it came the week before Christmas. Arriving home from work, I read the letter while standing at the mailbox, stuck it in my pocket, and rushed to Center City where Hania was attending a friend's holiday party. I walked into the apartment and found her standing by a lit Christmas tree. As she read the highly complimentary acceptance, my joy at my triumph merged with my delight at having Hania to share it with.

This success surpassed the one I had enjoyed with the *North American Review*. Apart from having to use a pseudonym there, I had been abroad when that piece appeared, which added to the slightly unreal nature of the experience. And the *American Scholar*, while also not a household name, had the distinction of being the journal of Phi Beta Kappa, which was. I thought of my father, fretting over my grades in high school, rejoicing when Jim made law review, seeing me now in the publication of the esteemed honor society.

The previous year, Theroux had published a collection of essays, *Sunrise with Seamonsters*, that included one on V. S. Naipaul. The two had met at Makerere University in Kampala, Uganda, where Theroux was a lecturer and Naipaul a guest writer. Theroux showed him something he'd written, and Naipaul, after offering a few editing suggestions, said he should publish it. "Send it to a good magazine—forget these little magazines," Naipaul instructed him. "Don't be a 'little-magazine' person."

I was becoming a "little-magazine person," but I didn't mind. At least I was appearing somewhere. Peter Meinke, back in Warsaw, had said that one of the great things about America was all the wonderful little magazines. Of course, he was a poet, with few other outlets. But "little-magazine person" sounded better than "ACP *Observer* staff writer." Or at least "contributor to the *American Scholar*" did.

The very existence of *Sunrise with Seamonsters*—a collection of literary and travel essays—was an encouraging sign that negated any possible slight I might have felt from something written inside it. Books about travel were now filling American bookstores the way they had always populated English ones; the great British traditions of writing and reading about the world had come to the United States. This was in large part due to Theroux and the commercial and critical success of *The Great Railway Bazaar* and also to Chatwin, whose *In Patagonia* had been hailed as a masterpiece by people who worked in more "respectable" genres. All of a sudden, publishers fell in love with travel books; some—Random House, Atlantic Monthly Press, Prentice Hall—started imprints devoted solely to them. M. F. K. Fisher's *Two Towns in Provence* was reprinted, and Leigh Fermor came out with *Between the Woods and the Water*, the radiant second volume in his series about his walk across Europe. Jan Morris's idiosyncratic essays on place were appearing regularly in *Rolling Stone*. All of this was exhilarating for someone whose goal was to become a travel writer, who was at that moment writing a travel book. But it also made sense: Travel writing's long appeal in England had grown out of that country's history of empire, and now the United States was the dominant world power. It seemed natural that Americans would want to learn about the places over which their country wielded influence.

Brits like Chatwin were showing us the way; in the case of Theroux, and his occasional hauteur, some of the old practitioners stood out as role models. One weekend, visiting Hania's recently arrived cousin in New Jersey, I entered a bookstore and discovered *Granta*. It looked more like a widened paperback book than a magazine, especially with the familiar logo of the penguin standing attentively in its orange oval. On the cover was a close-up of a tribesman with an extended lower lip, above the words "IN TROUBLE AGAIN: A Special Issue of Travel Writing."

The table of contents contained a mix of the familiar and the unknown, the old and the new: Salman Rushdie, Colin Thubron, Timothy Garton Ash, Amitav Ghosh, Martha Gellhorn, Hanif Kureishi, Norman Lewis, Ryszard Kapuściński. Now, back in New Jersey, I was finally reading the work of the Pole who had eluded me in Warsaw—though on Angola, not Poland. The cover story was by a man named Redmond O'Hanlon, whose gift for comedy, especially when describing the pain certain creatures in the Amazon can inflict, won me over to adventure tales.

I tracked down earlier issues in used bookstores. *Granta* had started life as a student magazine at Cambridge University and now, under the editorship of an American, Bill Buford, it had become travel writing's unofficial house organ. This late twentieth-century flowering of the genre was reminiscent of the between-the-wars period that Paul Fussell had examined—O'Hanlon, Thubron, and Kapuściński replacing Fleming, Byron, and Greene—and a rebuke to his pronouncement, at the end of that book, on the death of travel writing, a genre that, he claimed, could no longer exist now that the age of travel had given way to the age of tourism.

The new age was ushering in a new kind of travel writing. People had been penning tales of their dangerous expeditions for centuries but not from such mock-heroic stances. Kinglake in *Eothen* had shown sangfroid arriving in Cairo during the plague, but he hadn't been self-deprecating like O'Hanlon. And he used a classical term for his book title, reflecting a more literate time, not the half-bemoaning, half-boasting "In Trouble Again," which befit our more exhibitionist era.

Travel writers were no longer retreating from the scene in their books, letting the locals and their environs speak for themselves; they were the main characters in nonfictional picaresques. They took Waugh's first-person junkets to a higher, more plot-driven level. In *Old Glory*, published in 1981, the British writer Jonathan Raban sailed the length of the Mississippi River, capturing memorable people and moments but also telling of his personal journey—an adult, solitary, immigrant Huck Finn, whose downriver progress was momentarily halted by an affair in St. Louis. Like Theroux, he was infusing and enriching the travel book with elements from the novel, not the least of which were narrative arc and engaging protagonist. Readers could eagerly follow the tale of the author's passage while, almost subliminally, learning about the lands he passed through.

Unlike Theroux, Raban brought a foreign eye to familiar places, which was also a feature of some of the new travel writing. In a world that was increasingly being visited by tourists, he went where the tourists lived, in this case the small towns and prosaic cities along the Mississippi. And, using his deft analytical skills—aided by a formidable knowledge of history and culture, geography and religion—he was able to make readers see them afresh. Interpreting a landscape, wresting out its meanings as opposed to simply describing its features, was another aspect of the new travel writing, one essential with the growing ubiquity of the camera.

For travel writers it was a heady time. And, as once with Solidarity, I saw no reason why it shouldn't continue.

There was a writers' group in Philadelphia that met occasionally at the Philopatrian Literary Institute, which occupied a building on Walnut Street as resplendent as its name. Carlin Romano, the *Inquirer*'s book critic, sometimes dropped by, as did Ben Yagoda, then an editor at *Philadelphia* magazine. I wrote a few reviews for the former, but never had any luck with the latter. City magazines had already started their sad decline into "10 Best Hoagies" and "100 Best Doctors."

One Sunday morning a few blocks away, driving through the quiet streets on my way to St. Peter's Church, I got slammed in the middle of an intersection. My Datsun skittered on its two left wheels, almost flipping over, and came to rest inches from a lamppost on the opposite curb. The man who had hit me stopped and got out and, after verifying that I was unharmed, informed me that I had run a red light. I told him that my light had been green. No, he insisted, his had been green. There were no witnesses; it was my word against his, and he was a lawyer, and I was in shock.

I managed to rumble back to West Philadelphia, but the damage to the body made the car unsalvageable. A friend, observing the passenger side, remarked that I'd been "T-boned." I remembered that S. J. Perelman had found the experience of crashing his sports car slightly less lamentable because it had taught him a new word: "totaled." I had never heard "T-boned"—a marvelously descriptive verb—but it was not worth losing my first car to increase my vocabulary. I had been doing that more effectively and much less traumatically by reading.

We quickly got used to life without a car. I walked to work, as always, and we took the bus to Center City. Shopping by means of public transportation was a bit of a hassle—lugging our Amish chickens and Jersey tomatoes home from Reading Terminal Market—but we were used to it from living in Warsaw. And not having a car meant I got more writing done.

But it did seem to symbolize the stasis my life had fallen into. One day I received a letter on *New Yorker* stationery:

Dear Mr. Swick,

The letter of last August, together with the fine article on Poland you wrote for The American Scholar, was unfortunately misplaced and just now comes to my attention. Please forgive me for not replying sooner. I am no longer The New Yorker's editor-in-chief. The new editor-in-chief is Robert Gottlieb. You may wish to get in touch with Mr. Gottlieb.

Best wishes,
William Shawn

His signature was minute, as if also apologetic. I saved rejections—like war wounds, they were a kind of badge of honor; this one I framed.

I doubted Mr. Gottlieb would be any more receptive than Mr. Shawn had been. And writing about doctors was getting me nowhere. At lunchtime I'd wander over to the Penn campus, buzzing with young people buoyed by bright futures. One afternoon I came upon a used clothing sale—two small racks set up next to a walkway—and was held there by an achingly beautiful tie. It was thin, with horizontal stripes in a repeated pattern of black, olive, beige, and turquoise. In keeping with the horizontal theme, it ended not in a point but in a flat bottom. Turning it over I saw the label, the word "rooster" written in uneven red letters next to a black-and-red caricature of the barnyard bird. I had not known about "rooster ties," nor was I aware that, in future years, they would be the only tie I'd wear, choosing from a collection of more than fifty.

New neckwear improved my mood only slightly. I missed Vernon, who had died the previous August. He had begun experiencing excruciating pain throughout his body; he claimed it was acute osteoporosis. I would go to his apartment after work and find him naked on the bed, writhing in agony in his half-furnished room. The Lonely Death of the Freelance Writer. I felt helpless and, selfishly, wished to be elsewhere, walking the streets, enjoying the long days. Summer, if not my life, was just beginning.

Vernon eventually was admitted to Graduate Hospital, where his suffering was partially relieved by drugs. Visiting one day, I found him watching a Phillies game.

"Since when do you like baseball?" I asked.

"I like precision," he said. "And this man"—he nodded to the pitcher—"is very precise."

He died a few weeks later and was cremated. An obituary appeared in the *New York Times*—the work, I assumed, of Rebecca Sinkler—and Jerre and I and a female poet scattered his ashes in the Schuylkill River. Too far from the current, they dirtied the still water along the bank.

I inherited his books, including *his* books—*Cinema Borealis* and *On Film: Unpopular Essays on a Popular Art*. All of them reeked of cigarette smoke. Before sleep I heard his hopeful voice answering the phone and the fresh memory caused me to sob.

One day a call came from Phil. He had left the *Los Angeles Times* and come back east, first to Connecticut and then to Rhode Island, where he was now an editorial writer at the *Providence Journal*. There was an opening for another, and he wanted to know if I would be interested. Writing editorials was something I had no desire to do and, quite possibly, no talent for. I had a tendency to see both sides in an argument, a fact that often made it difficult for me to take a stand. To a certain extent this is a good trait in opinion writers, who need to consider both the pros and the cons, but eventually they have to decide one way or the other. Or at least their publisher does. And the publisher of the *Journal*, I knew from the fact that he had hired Phil, was deeply conservative. I was culturally conservative and politically liberal. I had rarely discussed politics with Phil, or anyone else; my interests lay elsewhere. The back of the book was my preferred half of the *Spectator* and of every other magazine that had the divide. The little-magazine writer was a back-of-the-book reader. My fondness for "arts & letters," periodicals' sideshow, carried all the way to the end-page essay, which was possibly my favorite form. Viewed by serious journalists as a trifling afterthought, it often had a timelessness that the cover stories and in-depth analyses almost never enjoyed. I got the impression from Phil that some of the editorials I'd write would be along these lines. The Gift of Thanksgiving. Opening Day at the Ballpark. Look what Updike had produced after he stopped by Fenway for a final glimpse of Ted Williams.

I flew to Providence for a round of interviews. My pro-Solidarity, anticommunist history impressed the publisher, who mistakenly assumed I could also write with authority on health-care issues. He wondered about my economic philosophy. So did I. I burbled something; he let it slide. He was decorous and soft-spoken, from an old WASP family; the man just beneath him, round-faced and horn-rimmed, came across as steely; I got the impression he saw right through me. It was a testament to their respect for Phil that I was offered the job.

I accepted. It was my dispirited path back into newspapers, my last chance to gain the crucial years of experience. Hania would stay in

Philadelphia, where she was now working, so I would be on my own in New England, doing a job I wasn't cut out for. I was falling again into an unexpected place but, this time, with an unfamiliar feeling of heaviness.

CHAPTER 7

Undivine Providence

WALKING TO MY FIRST DAY OF WORK AT THE *PROVIDENCE JOURNAL*, ON A
crisp September morning in 1987, I picked up the paper and read on the
front page that the publisher was in the hospital in critical condition.
According to the story, his unconscious body had been found early Sunday morning on a lonely country road next to his fallen bicycle.

When I reached the office, Phil and the other editorial writers were
grouped in solemn conversation. There was talk of foul play; the *Journal*
had run in-depth investigations of corruption and mob activities in the
state. We had an abbreviated and subdued editorial meeting, after which
I overheard one of the columnists, the lone liberal in the department, say,
"It's not a good time for Phil's friend to start."

Editorial was on a different floor than the newsroom, separated from
it physically as well as ideologically. Because of their politics, editorial
writers were not well viewed by reporters, the vast majority of whom, as
at most American newspapers, were liberal. I got a sense early on of the
disdain they held us in, which was a new feeling for me—not being disliked (I had been a freshman in high school) but being disliked for beliefs
I didn't necessarily hold. Of course, I could not admit this to people in
the newsroom without calling my job qualifications into question. So, as
a result, like my colleagues, I generally kept my distance from them.

It would make for a lonely existence in Providence, a city that had
certain similarities to Trenton. It was the riparian capital of a small
state, with a prominent Italian population. Though Federal Hill, unlike
Chambersburg, had been prettified with old-fashioned streetlamps and

dolled-up restaurants. It also had more of a Mafia connection. Both Providence and Trenton were historic cities, but the Revolutionary-era buildings in Trenton tended to be military and tucked away, while Benefit Street was a handsome stretch of still-inhabited Colonial and Federal-style houses rimming arborous College Hill. Providence's downtown was more vibrant than Trenton's, thanks to the presence of the Rhode Island School of Design and, up on the hill, Brown University. Instead of a nearby Ivy League school, Providence had one in its city limits. This gave it even more liberals to hate me.

My situation here was completely different from what it had been in Trenton; no longer the wild card in the newsroom, I was now one of the pariahs in editorial. Often, when a local asked me where I worked, the look of interest would slowly turn to disfavor if I were pressed to reveal my position. Even at the Episcopal church near my apartment the priest's brown-toothed smile, on hearing of my new profession, disintegrated into an unchristian grimace.

This would have been bearable if I had had a job I loved, but, for the first time since Washington, I had one I hated going to every morning. I would arrive a good hour before the editorial meeting at ten and nervously scan the paper for ideas. Then, feeling inadequate, I'd walk into the conference room with my feeble defenses.

The meeting was led by the chief editorial writer, a large, bald, cigar-smoking, suspenders-wearing Irish American who had all of Phil's conservatism—in fact, a good deal more of it—and none of his sophistication. Foreign visitors always found him fascinating because he conformed so perfectly to the stereotype of the American male created by Hollywood in the 1950s. His gruff exterior concealed the proverbial good heart, though his kindness was reserved mainly for people who looked and thought as he did. Between puffs on his cigar, he would announce his views on the issues of the day and then we'd go around the table: the bearded, owlish young man who fancied himself the next William F. Buckley Jr.; the cheerfully cynical Italian American who had attended Boston Latin; the lone liberal; and Phil. I went next-to-last, before Phil, and said as little as possible. It wasn't just that I was more comfortable writing than speaking; I lacked confidence, usually, in my grasp of my subject and, eternally, in the support of my listeners. Occasionally Phil would jump in and redirect my wavering words into a path that everyone could find acceptable. Sitting pensively at the head of the table was the interim publisher, the man who saw through me.

Finally, the session would adjourn and I'd retreat to my spacious cubicle where, for the next few minutes—before beginning composition of my vital editorials—I'd regroup with relief at having survived another meeting. Then I'd ask myself the question I assumed only Yankee reticence was keeping everyone else from asking me: What are you doing here?

I was getting back into newspapers, but I had moved far from features and the dream of perpetuity. Editorials have a shorter life span than even news stories; researchers combing through old newspapers rarely look at editorial pages. A persuasive editorial can change government policy, but I had not taken up writing with any lofty goals of reform or betterment; my modest desire was to entertain people and open their eyes to the world. Which is why I read more Waugh and Nabokov than Zola and Steinbeck.

And writing editorials was more sedentary than working for the American College of Physicians had been. It was assumed that, like my colleagues, I was already imbued with knowledge and experience and had no need to go off on reporting junkets or fact-finding missions. My job was to pore over the information others had gathered and pronounce on it in a fashion that adhered to the paper's political leanings. I didn't get out of the building, let alone the city. I was as office-bound as a copyeditor.

My first efforts, in the eyes of my boss, were too reportorial and lacking in commentary.

The publisher died a week after his accident. Flags at the newspaper, the Biltmore Hotel, even the Rhode Island School of Design, flew at half-staff. People in editorial felt as if they'd lost a friend; I felt as if I'd lost an ally.

I had Phil, at least, for both these rolls. On Fridays—the busiest day of the week, when editorials had to be written for Saturday's, Sunday's, and Monday's papers—we'd walk to the greasy spoon across the street for bowls of thick New England clam chowder. (I loved going out for lunch, which seemed to me one of the great rewards for working in an office.) One day Phil looked up from his spoon and asked, "Do you think it significant that the only artwork in here is an illustration of the Heimlich maneuver?"

Apart from a professed fondness for airplane meals, Phil was indifferent to what he ate. It was partly the intellectual's dismissal of the sensual, though he put a lot of thought into his attire, daily changing his watchband so it would coordinate with the colors of his tie. But it was also an aspect of his nonconformity now that Americans were becoming interested in food. A downtown restaurant that had acquired a reputation for authentic Italian cuisine was pretty much ruined for me after Phil noted that its name, Al Forno, sounded like that of a porn star.

I admired how he not only went against the current but did so wittily. He christened the liberal columnist "the poet laureate of conventional wisdom." One afternoon, after finishing his editorials, he wandered into my cubicle, took a seat in the chair, and pulled down my copy of *The Enigma of Arrival*. Opening to a random page, he began reading in a thick Indian accent. (He was an excellent mimic.) After a few lines he put the book down and produced a long yawn. "When I can't sleep," he said, now in the tones of an actor in a commercial, "I take Naipaul." When I finished laughing, I told him to read *A House for Mr. Biswas*. Once, he spotted on my desk a book by Kazimierz Brandys that had recently been translated into English. "More Eastern Europeans," he asked, "sitting around bemoaning their fate?"

He claimed to hold a certain sympathetic fascination for General Jaruzelski, which was a result of his strict contrarian principles: Whoever was loathed by the majority was often admired by Phil—up to a point, and, as a fairly bloodless dictator, the general didn't surpass it—and whoever was adored by the masses was usually disliked by Phil. He held in affection public figures the rest of the population dismissed as boring: Eisenhower was one of his favorite presidents, and he had great admiration for Prince Charles. He provided a useful reminder to consider the reverse side of popular opinion, particularly if one didn't always agree with it. Even those times when he simply toed the party line, Phil showed me, because it wasn't my line, to look at things thoroughly and not just think like everyone else. His contempt for so-called received wisdom was an invaluable lesson, not least because it carried over into the realm of geography. In his newspaper career he had bounced around a fair amount—Anniston, Alabama; Lexington; Los Angeles—and discovered the merits of places outside the Eastern Seaboard. He hated how Hollywood, in movies and TV shows, invariably gave the unintelligent character a Southern accent. And he tirelessly pointed out the anachronisms—from hairstyles to figures of speech—that marred, at least for him, historical films. When I wrote about Lake Superior State University's annual list of banished words—the sort of apolitical editorial I couldn't get enough of—I questioned the authority of people living in "the middle of nowhere" (Sault Ste. Marie, Michigan). Phil, reading the editorial before publication, pointed out that the term was a matter of perspective and noted that Providence was hardly the center of the universe. My admiration for the unsung, I was made to realize, embraced only the places I knew; Phil taught me to extend it beyond personal experience.

He had a wife and two small children, so I rarely saw him outside the office. And on weekends Philadelphia often beckoned. Friday, after writing my last editorial, I would carry my bag to the Amtrak station to catch the Benjamin Franklin. While waiting, I would read the words of Robert Louis Stevenson that had been carved into the floor—"For my part, I travel not to go anywhere, but to go. I travel for travel's sake. The great affair is to move."—and be reminded that I had never traveled like that. There had always been a reason for my movement that outweighed my love of motion.

On good nights, the train would arrive in Providence at 6:40. I'd try to find an aisle seat—window seats were quickly taken—in the middle of the car, so as to be away from the regularly opening doors and the equally consistent drafts. My seatmate was usually a woman—I found women more intriguing than men—and very often a student. The Benjamin Franklin was also the Ivy League Local, starting in Boston (Harvard) and stopping in Providence (Brown), New Haven (Yale), New York City

(Columbia), and Princeton Junction before ending its run in Philadelphia (Penn). On Friday evenings, the median IQ of the Benjamin Franklin was greater than that of any other train in the country.

It made for interesting conversations, especially when the train stopped and the power went out, stilling highlighters and plunging everything into darkness. This happened on every journey in New Haven, when the train went dead on command, a diminishing expiration—PPPSSSSShhhhhhh—of lights, motors, and fans that preceded the replacement of the diesel engine for one that ran on electricity. Any Japanese tourists onboard must have found the procedure impossibly quaint.

After twenty minutes, if we were lucky, the train continued on its now-greener passage through Bridgeport, Stamford, New Rochelle, where I always thought of Rob and Laura Petrie (*The Dick Van Dyke Show* had first planted in my mind the idea of becoming a writer). Soon we were approaching Manhattan like Santa's sleigh, grazing the flat rooftops of neighborly Queens and, minutes later, burrowing beneath the island known for its heights. When the train came to its subterranean stop, most of the passengers grabbed their bags and departed, replaced by panhandlers with stories of tickets lost and money stolen week after cursed, unlucky week.

At about half past midnight, we'd arrive at 30th Street Station in Philadelphia, where a taxi took me to 47th and Larchwood and the bed made warm by Hania.

On one of these trips in October I went to the Philopatrian Literary Institute for a meeting that had been arranged with agents and publishers. There was a tall, unsmiling man from St. Martin's Press and a well-dressed, familiar-looking woman from St. Peter's Church: Barrie Van Dyck. She told me she had recently opened a literary agency with a friend and asked me to send along my Polish book.

A week later I got a call at the newspaper. Barrie was "savoring" my writing—the best, she claimed, her agency had seen—and was eager to take me on as a client. I told her I'd try to have the book finished by spring.

I was elated, even while realizing it might not lead to anything. But it gave me the incentive I needed to continue. There had been times when I had sat at my desk wondering if I were wasting my time. The clamor for travel writing had seemed very far away.

Now, after work, I walked the twenty minutes back to my apartment on Charles Street and there, at my desk overlooking the Moshassuck River, wrote about Poland. It was the same desk I had had in Phillipsburg and, though I now owned a word processor, I was still composing in longhand at it. When I finished a chapter, I'd type it into the word processor and then print it out. Sometimes I'd send excerpts off to magazines.

These joyless, disciplined days had a few echoes of Arta. To save money on heating, I slept in a cold bedroom, wearing the nightshirt and nightcap I'd bought as a joke in a vintage clothing store near the Rhode Island School of Design. And I spent much more time writing than talking—the productivity of the friendless. But the writing, now, was taking the shape of a book, giving me some hope of a reward at the end. And Greece—for all its misery—had been nothing more than a brief, misbegotten foreign episode; Providence was wage-earning, career-settling, soul-wearying life.

In Philadelphia, I wallowed in weekends that, ultimately, accentuated the universal melancholia of Sunday evening. I would just get used to waking up next to Hania, watching her drink her tea, hearing her endearments, and then I would have to pack and depart. I had thought we were through with separations, but the long, distant ones that had pockmarked the early years of our relationship had balefully evolved into a series of short, niggling ones.

With a bag from the local gourmet shop, I'd board the Merchants Limited, a name perfectly chosen for the passengers contained inside, people who resembled—who sometimes *were*—hungover versions of the ones who had taken the train on Friday. I wasn't the only person headed back to the grind, but I was one of the few who would feel like an imposter once he got there.

Some Sundays I'd arrive at my apartment and find a rejection letter to add to the gloom.

My first semester at Villanova, after taking the last of my final exams, I treated myself to a movie in the nearby town of Wayne. The film was *The Owl and the Pussycat*, and the leading male character was a struggling writer. At one point he opened his mailbox and pulled out a letter of rejection. It was a brief, perfunctory scene—to demonstrate his lack of success—yet it's the only one from the movie I remember. A college freshman with literary dreams, I watched it with romantic yearning. To me, the rejection was a coveted sign, estimable proof, that the man was the thing I wanted to be: a writer. I couldn't wait for the day when I got rejections. Today, when many editors simply ignore unwanted submissions, the psychological value of a response is more apparent. A rejection, however demoralizing, is nevertheless an affirmation, an acknowledgment that you are a writer who has written something. A non-answer seems to suggest your non-existence.

Are neurotics naturally drawn to writing, or does a life spent depending on the approval of others simply make one neurotic?

A few weeks before Christmas I came home from work and found a letter from *Ploughshares* in my mailbox. Opening it up, I read that the poet Philip Levine, who was guest editing the next issue, had accepted my "Martial Law Journal." Not only would *I* appear in the quarterly, but so would a few of my students, through their letters.

The difficulty of baseball is often illustrated by the fact that the best hitters fail two-thirds of the time. Freelancers have far worse percentages. We go months without a hit, enduring slumps that would end the career of any major leaguer. But we don't keep averages for the simple reason that one acceptance negates every rejection. It is the great beauty, the saving grace of writing: that all you need is one editor, or one publisher, to say yes, and then everything that came before is rendered immaterial. An acceptance is not just a grand slam, it is a grand slam that, miraculously, erases every strikeout and ground out and pop-up that preceded it.

Though, since I was clearly still a little-magazine person, perhaps this one was just a home run.

In the new year, I spent every weekday evening writing my book, which I had given the working title of "Polish Days"—a nod to Tuwim's *Polish*

Flowers. Walking through snow to and from work helped me recall my winters in Warsaw.

By spring I was finished. I printed out the final pages while the theme from *Hoosiers* blared from the TV. I saw it as an encouraging sign.

The following weekend I took the train to Trenton for a newspaper reunion. There was a modest turnout at a historic tavern that Sally and Sam had recently purchased. One former reporter, a woman no longer in journalism, asked if I was still writing. The fate of the little-magazine person.

Mark Jaffe gave me a ride to Philadelphia, where he was now working at the *Inquirer*, and on the way I complained about my lot.

"I don't like writing about issues," I told him.

"What *do* you like writing about," he asked, seemingly reasonably, "*non*-issues?"

"People and places," I said, as if that were a thing.

It was. The following day, walking near Rittenhouse Square, Hania and I stepped into Banana Republic. While she looked at clothes, I headed back to the book cove—every Banana Republic store had one—and found, amidst the guidebooks and travel narratives, a magazine called *Trips*. The cover carried a winsome illustration of a beach containing a pianist, an elephant, a nude reclining under a palm tree, and a bespectacled man in a black cap and blue suit turning his back to the curious assembly and the black-hulled ocean liner sailing in the background. Among the stories listed on the left were "Black Humor on the Black Sea" and "Richard Ford's Little Rock." Printed in the same black type, on the right, were the words "INAUGURAL ISSUE." It was obvious from the titles, not to mention the illustration, that this was not another glossy. Opening it up, I found more charming drawings—of Sofia and Oakland—some vintage photographs (Ford's piece was a reminiscence of growing up in his grandfather's hotel), and artful photographs of stockmen in Australia's Outback. Turning to the front, I found a letter from the editor-in-chief, Mel Ziegler, who was also the founder, with his wife, Patricia, of Banana Republic. That a clothing company was now producing a travel magazine—to be sold in its travel bookstores—seemed further proof of my desired profession's unfettered ascent. *Trips* took its spirit and tone from *Holiday*, and, though it appeared to have come out of California, there was an obvious appropriateness to my finding it in Philadelphia. I carried it excitedly up to the cashier.

On weekends when I didn't go to Philadelphia I'd often take the bus to Boston. I had pretty much given up on Rhode Island. In many of its quarters—at least the ones in which I would have liked to be—I was persona non grata. But the place seemed difficult even for newcomers who didn't write editorials. The state made New Jersey seem big, and not just geographically. I was in New England, I knew, no longer in the friendly confines of the Mid-Atlantic, but the Ocean State struck me as a place apart, possessing its own unique character. For New England, it was quite ethnic, with its Italians and Portuguese and French Canadians, all people generally known for their warmth. But because of the close borders or, perhaps, the pervasive mentality of the region, an entrenched cliquishness prevailed. Everyone had their friends from childhood; families that had lived for generations in the same towns went every summer to the same beaches. There was not a lot of interest, I got the impression, in adding to the unit. It felt as if I'd moved not to another state but to another country. I spoke the language but I didn't know the customs. And I probably didn't belong.

I didn't really take to Boston either, though it was a metropolis, a place in which one could lose oneself. The Boston Latin grad, illustrating the difference one day between Providence and Boston (or any great city), said, "In Boston, you can turn a corner and fall in love." I recognized its red-brick similarities to Philadelphia—Beacon Hill and Society Hill, the North End and South Philly—and acknowledged that it was the superior city. Unlike Philadelphia, which was financially and culturally eclipsed by New York, Boston stood alone as a regional capital. This only deepened my affection for Philly, as did the fact that Boston was a much more segregated city. It was cold and blustery for long periods of the year, overrun by young people in scarves and backpacks. Faneuil Hall Marketplace was a Rouse Company property, populated by chains, with none of the Italo-Amish soulfulness of Reading Terminal Market. Watching the patrician women in Copley Place, polished and unapproachable, I would remind myself that Hania had a more venerable lineage.

But the city offered attractive cultural events. One week in June I went up on a Monday evening to hear Paul Theroux and returned on Wednesday to see Eric Newby. Theroux spoke at the Women's City Club on Beacon Street about his new book, *Riding the Iron Rooster*. At the signing afterward, he sat attentively on the edge of his chair, his back very straight, his thinning hair still very black. I handed him my copy of *Sunrise with Seamonsters* and told him that I appreciated his comments, in the new book, on the beauty of Polish women. (To get to China he had taken trains through Europe, the Soviet Union, and Mongolia).

"Yes, well, they are extraordinary," he said in the accent he'd acquired while living in England.

"I know," I said. "I married one."

"Did you?" he said. "Good for you."

If there hadn't been a line behind me, I would have told him of the role his friend Pritchett had played in our union.

Newby appeared, with his wife, Wanda, at the Globe Corner Bookstore downtown. After his talk he sat in an armchair, looking rather frail. I leaned over and told him that, in Philadelphia, whenever I passed the *Moshulu*, I thought of him. He raised himself and, eyes bright with interest, grabbed hold of my arm. He asked me what it looked like now, and as I described the interior—the romantic restaurant, the low-ceilinged bar—his look of enthusiasm faded a bit. I imagined it was difficult to think of people enjoying date nights on the vessel that had made him not just a travel writer but also a man. He had gone to sea on the barque at the age of eighteen.

These outings brightened considerably my dim life in Providence. My bedroom was no longer cold but it was still dark, the curtains eternally closed against the traffic on Charles Street. (The apartment had dual personalities: One end faced a busy thoroughfare, the other a burbling river that resembled a stream.) I still walked to work with a feeling of dread, still suffered through the morning editorial meetings, still wrote sentences that gave me little satisfaction. I had never wanted to write anonymously—I was still irked that my first travel story had appeared under a pseudonym—yet I was never so grateful for anonymity. My parents came to visit one weekend and, when I showed them a few of my recent editorials—on Medicare reimbursement for Rhode Island hospitals, EPA reassessments of the cancer risks of toxins—even my mother had difficulty dredging up praise.

The highlight of every day was getting the next day's editorial and op-ed pages to proof, because there'd often be a cartoon by Pat Oliphant. Unlike us, Oliphant didn't take sides; every politician of every stripe was equally despicable in his eyes. His drawings were brilliantly conceived and executed with inspired, unsparing exaggeration. One day I read a profile of him in which the writer arrived at his home just after he'd finished a drawing. The cartoonist, with the gusto that comes from completing a good morning's work, suggested to the writer that they go out for lunch, lunch being one of his favorite things.

Sometimes my proofing would be interrupted by a call from Barrie, with news of the latest book rejection. A number of them—including the one from St. Martin's Press—complained that the writing was too impersonal. Many editors wanted more about Hania and me in the book; readers, they said, would be interested in hearing about our love story.

That spring, Houghton Mifflin—Theroux's former (and future) house—had published *Nothing to Declare: Memoirs of a Woman Traveling Alone*, by Mary Morris. It was the first of the introspective travel books, a subgenre I was philosophically opposed to at the time. Even when travel writers became the main characters—as Theroux and Raban did—their books were primarily about the place. What we learned about them was always secondary. This, to me, was what defined a travel book; as soon as the author took center stage, it became a memoir. And a memoir of a romance was the last thing someone should have been writing about Poland, a country that for two years had tried to alter history. Not to mention—and I didn't to anyone—that the love story would have been bogus.

I started to have a few successes outside the world of little, in less prestigious but more widely read and better paying magazines. One was an essay on learning languages that appeared in *Travel + Leisure*, the glossy I had long disparaged. It was not its typical fare—I suspected its editors felt some competition from the new and more sophisticated *Condé Nast*

Traveler—and today it is impossible to imagine such a piece running in either publication.

Rebecca Sinkler occasionally sent me books to review, giving me my first *New York Times* byline. Most were historical or political books on Poland, a familiar pattern that didn't bother me as much now since it illustrated, I hoped, the need for a different approach to the country.

She also put me in touch with the editor of the Travel section. On July 3rd, my paean to the Old World charms of Eastern Europe ran as that week's end-page essay. One month later, the section published "A Traveler's Finest Hour," about my experience selling *koulouria* at the Athens bus station. Meanwhile, *Nowy Dziennik*, the Polish newspaper in New York, reprinted the first essay, giving me the thrill of seeing my words in translation. I immediately became the writer in Polish I had the least trouble reading.

That summer I started preparing for our fall trip to Poland. My intention this time was to write about the problem of emigration, as the country's stagnation was producing waves of people seeking better lives abroad. It was an old, sad story—Mickiewicz, Chopin, Maria Skłodowska-Curie, Samuel Goldwyn—that I in a small way had contributed to. It was estimated that, since 1980, 600,000 Poles had left Poland, 265,000 of them permanently. In a five-year period in the middle of the decade, more than 21,000 people with a college education had left for good. "That is more," noted a former government spokesperson, "than a few developing countries possess altogether." I read Joseph Conrad, hoping to find his thoughts on the subject, and, to brush up on my Polish, reread the screenplays of Krzysztof Zanussi.

Our plane landed at Okęcie Airport on an overcast afternoon in September. As always, small groups of family and friends—some holding flowers—stood on the platform atop the small terminal. The sight never failed to move me; likewise the tears in the arrivals hall. People returning from what nearly amounted to forced exile had understandably high emotions, and there was the added poignancy of relatives embracing who now lived in separate worlds.

Over the course of two weeks I visited family—we stayed with Jaś and Elżbieta in their new house outside Warsaw—friends, old colleagues, former students, even my erstwhile tailor. The talk was always about life in Poland, conversations that often, even without my prompting, led to the subject of emigration. I interviewed a journalist who, giving the topic an interesting twist, had written an article for the Catholic weekly *Tygodnik Powszechny* on the people who *didn't* emigrate. I took the train to Krakow to visit the Institute of Polonia Studies, Polonia being the word that is used to identify the Polish community abroad. The fact that this group had its own name testified to its long and significant history. I read articles in newspapers and excerpts from the white paper on emigration from the Cardinal's Social Council. I seemed to go from depressing meeting to distressing report to doleful tête-à-tête, and the experience would have made my life in Providence seem almost rosy if not for the fact that I felt engaged and fulfilled in a way I never did writing editorials. I was doing what I loved—talking to people, hearing their stories—what I still dreamed would someday be my life's work.

We arrived back at Kennedy Airport on a Sunday evening. No one was there to greet us. On the bus to Philadelphia, Hania slept with her head in my lap. We took a taxi to our apartment, where she got ready for bed and I grabbed a chocolate to take on the night train to Providence. Of our now-countless partings, this was one of the most onerous. I desperately wanted to stay, crawl under the sheets and sleep with my wife. I was physically exhausted from the journey and mentally unready for my return to dissembling.

The train arrived early in the morning. Back at my apartment, I showered and dressed in jacket and tie. Then I stepped outside and walked toward the office. Halfway there I stopped, my head awash in self-pity and half-hearted rebellion. I turned and took a few unconvincing steps in the direction of Charles Street; then I righted myself and slowly, resentfully, continued my sad commute.

At the end of November my third end-page essay—an abridged history of my lifelong love of travel—appeared in the *New York Times* Travel section. My mother called with the news that Nancy Stableford, at Trinity Church, had told her how much she had enjoyed the piece. I was tempted to tell my mother to call her and ask her to put in a good word for me with her brother at the *New Yorker*.

The essay appeared one week before I received a rejection from our Sunday magazine. It came through the office messaging system, reminding me of the time someone in the features department had returned a magazine I had lent him in an interoffice envelope, with no note of comment or even thanks inside. I had wondered then, as I wondered now, if this was normal New England procedure, or if editorial writers were so odious that we discouraged actual personal interactions.

The magazine editor had requested the essay, saying it would be interesting to hear from someone who was living without a car. In his rejection, he noted that the piece was too rambling and that the humor got in the way of many of my points. (He was from the South, so I couldn't blame regionalism for his humorlessness.) He helpfully instructed me as to what an essay should be and said that he'd be open to considering a revision. After, evidently, I learned the art of essay writing.

Editors, I was discovering, are as varied as people from distant time zones or different historical periods. Each has his own tastes, her own definition of quality. Yet there is a hierarchy in the world of writing, as in all professions, and having enjoyed some success at the top of the journalism wing—at the nation's leading newspaper—I was abashed by my failure at my own middling journal. I spent a good while thinking of an appropriate response, choosing finally, in fine New England tradition, to remain silent. Though I eventually rewrote the essay slightly because I wanted people in the newsroom who didn't read the *Times* Travel section to see what I was capable of. This backfired miserably as the edits inflicted on it were so unrelenting and to me inexplicable that they seemed almost retaliatory.

One evening after dinner I walked to Angell Street to meet Edward Hoagland. Phil had told me he was guest teaching at Brown, and I had written him a letter, care of the English department, to which he had graciously responded. I had always enjoyed reading his work, even when his muscular, meandering sentences proved a little daunting and difficult to follow. He was solidly in the camp of prolix American writers, but he was that rare type who could write as passionately and knowledgeably about city life as he could about the natural world.

He opened the door of the house, a handsome man in a red sweat-shirt and black corduroy pants. I handed him the bottle of Bulgarian wine I had brought and followed him through to the dining room. His dinner sat on the table: a thick, gray piece of meat on a bone. He asked if I had eaten; I would have said yes even if I hadn't. He proceeded to vigorously cut the meat from the bone and, through his stutter, ask me questions.

I had generally found journalists easier to talk to than writers; the former are curious by nature, while the latter, especially novelists, are sometimes self-involved and seemingly oblivious to what's around them. But Hoagland was a novelist-cum-essayist with a travel bent and, as such, was a wide-ranging generalist. He expressed interest in the news-paper, local news, the book I told him I had recently finished. When I mentioned the responses I had been getting from publishers, he advised, "P-P-P-Put more of yourself in it."

He kept his eyes closed while struggling with a word, which was considerate as it relieved you from worrying about what kind of facial expression you should assume while waiting patiently for his utterance. I did wonder how he managed to teach. He spoke the most fluently when talking about himself. I hadn't realized that his first book, *Cat Man*, was published when he was twenty-three. The next two books, he said, got little or negative notice, and it wasn't until his late thirties—where I now was—that he gained some notoriety for his essays. I expressed my admiration for his ability to make a living from his writing, and he set me straight by telling of the teaching jobs he took out of necessity.

He spoke fondly of the days when he would take an essay to the *Village Voice* on Wednesday, see it in print on Thursday, and then get compli-ments on it at parties that weekend. He told me, in the voice of someone providing a lead, that a group of people who had left the *New Yorker* after the departure of Shawn were planning on putting out a new magazine called *Wigwag*. He mentioned that he had been in the same class at Harvard as Updike but didn't know him then; they became friends at the twenty-fifth reunion, when "the sense of rivalry had been eroded."

I was surprised by how forthcoming he was. He confided that he and his wife were in the process of separating, and he wasn't sure where he was going to live. He admitted that reading the travel magazines was discouraging, and if he did anything for them he wanted at least to get a good trip out of it. Yet they were reluctant to send him abroad, and domestic travel no longer enticed him.

"I've b-b-been lonely so much of my life," he said. "I d-d-don't mean that I've been alone—I've been loved—but as a writer you spend so much time alone, writing. I d-d-don't want to spend any more lonely nights in a motel in Louisiana.

"I'm older now. W-w-when you're in your fifties, the women you meet are forty, forty-five. I have nothing against older women. I like them. But there are fewer of them around. W-w-women that age are all attached. When I was in my thirties, the women I met were in their twenties. Then I loved traveling. I loved going to Louisiana. I had love affairs. I didn't go to Louisiana because I loved talking to trappers and game wardens. I had love affairs." His face brightened at the memory.

"Th-th-th-the same with Alaska."

He insisted that he was far from unique. Travel writers wandered; it's what they did. An exception, he said, was Colin Thubron. "P-p-perhaps that's why he gets such good insights."

At around ten thirty he asked if I would read a piece of his that had recently been rejected. The universal cut; what right did I have to complain? He led me up to his rented room; books he had written lined a shelf—in a neat row, as if in a bookstore—and a coyote skin covered the bed. He sat me in his desk chair and handed me the manuscript that had been keeping the typewriter company. The lines were tightly spaced; the pages numbered nineteen. I made my way as best I could—I'd drunk a bit of wine—through dense descriptions of the Wyoming landscape. It was the diary of a trip he'd taken with his daughter. I tried to think critically, but one part of my brain refused to engage, repeating stupidly but inevitably: I'm in Edward Hoagland's room reading an unpublished essay by Edward Hoagland. When I finished, I offered a few minor suggestions, for which he thanked me. Then I thanked him for the interesting conversation and headed off to Charles Street.

On Christmas Day, my first review of a work of fiction appeared in the *New York Times Book Review*. It was a novel set in Yugoslavia, written by a woman named Nadja Tesich. The author liked my review so much—I had called the main character "a female Holden Caulfield of the Balkans"—that she invited me to a party the following month.

The last year of the decade I never thought of as materialistic began with the inauguration of George Herbert Walker Bush as president—and the idea of "a kinder, gentler America."

One week later I left work early and boarded the train to New York City. I had told the *Times* Travel section editor that I was coming to town, and she had suggested I stop by so we could meet. Her office was on the eighth floor of the *Times* building, and I rode the elevator with a feeling of awe, even though I still found every page but one of the Travel section unreadable.

We had a pleasant chat, during which I got no indication that Nora was soon going to be transferred to another department. At one point, Bruce Chatwin's name came up; he was then suffering from a mysterious illness that he claimed was a rare fungal disease he'd picked up in China; Nora expressed concern that he might have acquired it while on assignment for her. (One year later he would die of AIDS.) Before I left, I made a self-deprecating comment about my lack of experience as a traveler—especially when compared to someone like Chatwin—and she reassured me by saying it was my "sensibilities" that mattered. In Trenton, Sally had sometimes used the word "Swickian" to describe my humorous pieces, but no one had ever spoken of my sensibilities. And now I was hearing about them from an editor at the *New York Times*. It was as gratifying as Hania's neologism for things she saw as imbued with my essence.

Back out on the street, I walked in the gathering darkness toward Central Park West. Winter's short days are further abbreviated by Manhattan's tall towers.

Nadja greeted me at the door of the apartment, expressing surprise that I had actually come. She had blond hair and a smile that disbanded shortly after formation. I wondered if this was the product of a Serb's tragic view of life or a tick she had acquired while living in dyspeptic New York. In addition to books, she had written a screenplay with Eric Rohmer that became the short film *Nadja à Paris*, directed by Rohmer and starring herself. Her brother Steve, who stood across the room, wrote the screenplay for *Breaking Away*. She introduced us, and I told him how much I had enjoyed the movie. (Not surprisingly, I had been drawn to the story of a small-town American teenager who pretends he's Italian.) Tesich had an un-Fellini-esque openness, and we talked for a good while. When he learned that I was working in Providence, he told me he had lived for a time in Saunderstown, and we happily launched into complaints about Rhode Island. "They are so obsessed with keeping up traditions," he said, "but they have no traditions to keep up."

I talked to the editor of *Commonweal*, who remembered my Polish Easter piece and wondered if I'd like to write about Romania. I confessed that I had only passed through the country on a train, nearly a decade ago, but that didn't seem to bother her. Such was the uniform view of Eastern Europe, even among East Coast intellectuals, that having lived in just one of the countries seemed to qualify you to write about any of them. I was also struck by the fact that, by attending a party in Manhattan, I had gotten an assignment. Writing, clearly, was a game of connections, and the thought occurred to me that I should perhaps overcome my disadvantageous love of the underdog and move to New York.

The visit inevitably highlighted the dullness of Providence. Evenings now, I spent writing my essay about Polish emigration. I had a lot of material: articles, studies, and a full stenographer's notebook that, thankfully, had made it through customs. To set the tone, I began the piece with a detailed and, I hoped, moving description of arrival at Okęcie Airport. You can't have reunions without departures.

When it was finished, it ran to fifty pages. I sent it to the *New Yorker*, the only magazine I knew that ran pieces that long. Of the various things I had submitted over the years, this was the one I thought the worthiest.

Ploughshares appeared, with my "Martial Law Journal." I heard from a few of the friends I sent it to—"Congratulations on getting at least one month published," Andy wrote from Chicago, where Mercedes was working in the Mexican consulate—but no one else. A writer creates in solitude and, very often, publishes to silence.

In April, Hania made a rare visit to Providence. I picked her up at the train station on Friday afternoon—I had rented a car for the weekend— and then took her to the apartment to get ready for dinner. My plan was for a romantic evening on Federal Hill. Before entering, I opened the mailbox and saw a thick manila envelope. My address was written in my hand; the return address carried the name of the *New Yorker*.

I was crushed. More than a rejection, it felt like a verdict, a decision on my fate. An appearance in the *New Yorker* I believed, probably wrongly, would be my ticket out of Providence, which I wanted just as badly, if not more so. Adding to the pain of the rejection was the timing. (I didn't realize that, twenty-five years hence, I would receive rejections while *eating* dinner.) That evening, and the rest of the weekend, I merely went through the motions; nothing could assuage my immense disappointment. And in my grief—for that's what it felt like—I made Hania feel not only awful but helpless. When Sunday evening arrived, I was relieved—for both of us—to see her off on the train to Philadelphia.

Because of my scorned vocation, I didn't spend a lot of time at Brown. Every few weeks I'd walk to the College Hill Bookstore to scan the magazines and smile at the saleswoman from the Azores. On good evenings, I'd walk out with the new issue of *Granta*.

One was built around the theme of "Home," demonstrating that travel writing is writing about place and doesn't necessarily involve transportation. Inside it I found Jonathan Raban on the cyclone that had raged through London, Norman Lewis on Essex, Martha Gellhorn on the Thirties, and another American, a man by the name of Bill Bryson, on Iowa. I read this piece with interest—talk about unsung—and was disappointed by its glib exaggeration and mocking humor. Titled "Fat Girls in Des Moines," it stood out from the neighboring works of reportage and proper memoir and made its author, an ex-Iowan living in England, sound ingratiating. One of the easiest ways for an American to score points with the English is to make fun of Americans. I knew this from years of reading the *Spectator* and now, under Phil's tutelage, I found it distasteful.

Travel writing was so big it was attracting people from other genres. P. J. O'Rourke wrote a book, *Holidays in Hell*, about his visits to the world's trouble spots. I picked it up in a bookstore one day and turned doubtfully to the chapter on Poland. Then I stood in the aisle, awed and humbled, for the next thirty minutes. O'Rourke had dropped into

Poland, without a word of Polish, and perfectly captured the spirit of the place, at least as a recalcitrant socialist state. He described things that for me, after two-and-a-half years, had become commonplace but now, presented through his undulled eyes, appeared afresh in all their absurdity. He was funny, of course, but with jokes that revealed truths while also provoking laughter. Some of them were Polish jokes.

A woman sends her husband out to the meat shop, where he takes his place in a long line. After six hours, the butcher emerges and announces that there's no more meat. The man is incensed.

"I am a worker!" he shouts. "I am a veteran! All my life I have fought and toiled for socialism! Now you tell me there's no more meat. This system is stupid! It's crazy! It stinks!"

A member of the secret police steps out of the line and approaches the man.

"Comrade," he says, "control yourself. You know what would have happened to you in the old days if you had talked like this?" And he turns his hand into the shape of a pistol and points at the enraged man's head.

Returning home empty-handed, the man is greeted at the door by his wife.

"What's the matter," she asks, "are they out of meat?"

"Worse than that," the man says. "They're out of bullets."

O'Rourke came to Brown to promote the book and, naturally, I went to hear him. As one of the nation's most liberal colleges, Brown was a bastion of political correctness even before the term entered everyday speech, and about a quarter of the audience consisted of protesters. A good number were Asian students, who apparently objected to his comments about Koreans. (South Korea was one of the countries covered in the book.) They took up whole rows, many of them with white T-shirts over their blouses and long-sleeved shirts; a few carried signs.

O'Rourke walked onto the stage and, standing at the lectern, began his talk. He paid no attention to the protesters, who were impossible to miss, with their signs now raised. Not wanting to infringe on his right to speak, they didn't heckle or harangue; they just sat quietly in pockets of shared hurt. Then, after about fifteen minutes, they stood up, held hands, and, row by row, silently vacated the hall. When their exodus was about two-thirds complete, O'Rourke halted his presentation for the first time.

"They're cute," he said, looking out over the last of the retreating minions. "No, really, they are. In my day we would have burned the building down."

Later he explained that he didn't pick on particular nationalities; he told the audience to check and see what he had written in a forthcoming article about his own people—the Irish. He was, he said, "an equal-opportunity offender." The non-graphic Oliphant.

A few days later, proofing the op-ed page, I noticed that the lines of type were strangely jumbled. In one word, one letter would sault above the others. I blinked, repeatedly, thinking perhaps I had something in my eye. But the dancing text persisted.

I called an optometrist who could see me right away. When I covered my right eye, the world returned to normal. When I covered my left eye, there was a big blur. The condition, the doctor believed, was histoplasmosis, but for confirmation and treatment I would need to see a specialist.

I called Hania, who sprang into action, getting me an appointment at Hahnemann University Hospital through the wife of a friend. Then I took a taxi to the station. It was the least pleasurable train ride since the Sunday of my return from Poland. I was in no mood for conversation, and, rather than attempt to read, I sat pondering what would happen to me if it turned out I couldn't.

Hahnemann stood on North Broad Street, not far from the milk-white *Inquirer* building. An ophthalmologist examined and then photographed my eyes. When he returned with the results, Hania asked immediately if the condition was treatable. When he said, "Let's talk," I knew it wasn't.

We followed him into a consultation room, where he proceeded to tell us about presumed ocular histoplasmosis. It was a disease believed to be the work of a soil fungus that caused abnormal blood vessels to grow beneath the retina. (So I, not Chatwin, had acquired the fungal disease.) These blood vessels formed a lesion that developed into scar tissue that replaced the normal retinal tissue in the macula, the central part of the retina. This explained why I still had peripheral vision in that eye. The bad news was that my condition, because of its location, could not be treated; the good news was that it usually occurs in only one eye. The ophthalmologist asked if I had spent time in the Ohio or Mississippi river valleys, as that was where "histo," as he now called it, was most prevalent. I told him no, just the Delaware River valley, working for three summers on its bridges. He said that it was also found in soil where bird droppings had accumulated, and I remembered the day I had spent shoveling pigeon droppings from a small chamber beneath the Phillipsburg free bridge.

The doctor gave me yellow-tinted aviator glasses to place over my regular glasses; they would help a bit, he said, in alleviating the distortions. He also provided us with the address of a store where I could buy an eye patch. His most encouraging words—especially since I cringed at the idea of wearing yellow aviators over my round tortoiseshells—were that the brain is an amazingly flexible organ that can make adjustments over time.

We went to the store to buy the eye patch, and I gallantly made a few pirate jokes. But I was very concerned, as was Hania. Reading—my love and my livelihood—now looked as if it were going to be a struggle, more a nagging frustration than a daily pleasure. With my new lack of depth perception, I worried about driving even though I didn't have a car. Everything now seemed compromised and uncertain.

I returned to Providence in very low spirits. My parents called, both saying they wished we could change places, that they could bear this disability instead of me. My brothers called and told me not to worry, that everything would be okay. Neither of them knew anything about the disease, but their loving, baseless assurances made me feel better.

At work, the secretary told me that she had gotten soap in her eye while taking a shower that morning and the pain of it had made her think of me. I told her I had experienced no pain, no discomfort whatsoever. That was the strange thing: Silently, stealthily, the condition had struck. One day I was reading and writing and putting the cap on my pen without a hitch, and then all of a sudden my world turned loopy, and I became maladroit. The only suffering was psychological, and it was ongoing.

I had only ever been able to close one eye—the left one—while keeping the other one open, and now I did it as little as possible. It was too frightening, even for a few seconds, to confront the great blur. When I had to cover it, at the doctor's, I could make out the **E** by looking away from the chart—thanks to my peripheral vision—while the other letters were indecipherable.

But slowly, gradually, I became less conscious of the problem; as the ophthalmologist had promised, my brain accommodated. I had already put the tinted glasses and the eye patch in a drawer, though I was sometimes tempted to wear the latter for sympathy purposes. I still occasionally missed the glass when pouring wine, and my right forefinger was perennially streaked with blue from my pen's failure to find its cap on the first try—making me the ink-stained wretch most of the people in the newsroom had long viewed me as. But I was starting, happily, to read with less difficulty. Though my eyes—and/or brain—seemed to tire more quickly.

Into this odd crucible arrived one day an issue of *Editor & Publisher* magazine. I always checked the classified ads in the back, and this afternoon I read that the *Fort Lauderdale News & Sun-Sentinel* was looking for a travel editor. It seemed a long shot, but I responded with a letter and some of my clips.

At the beginning of May, I received a letter from the *Fort Lauderdale News & Sun-Sentinel* features editor, along with five copies of the travel section to critique. I assumed I had the *New York Times* Travel section to thank. I took them to Philadelphia and spent most of the weekend poring over the articles and writing my comments. Now, no doubt because of the urgency, the histo seemed more of a hindrance again.

The sections, I wrote in my critique, were typical newspaper travel sections, telling of the sights and ignoring the locals. They skimmed the surface, providing information that could be found in any guidebook while paying no attention to everyday life. A good travel section, I wrote, should contain stories that people would want to read regardless of whether they were planning to travel; it should appeal to everyone who picked up the Sunday paper. And it would, I confidently if naively suggested, if the writing were evocative, insightful, surprising (like travel), funny (ditto); if it educated as well as entertained. My response was as unconventional as my *Trenton Times* application had been, but replacing the flippancy was ardor, with a probably misplaced note of proselytizing. It represented the philosophy of someone who had not backpacked but lived abroad. I believed that the travel writer's experience should approximate, as closely as possible, that of the expat; that this approach would produce insights unavailable to the dedicated sightseer. I wanted, perhaps futilely, to bring some of the elements of the travel writing that was appearing in books into the newspaper travel section.

About a week later, the features editor called and asked me to come down for an interview. I flew into Fort Lauderdale and took a taxi to the Riverside Hotel on Las Olas Boulevard. Was this, I wondered, where my dream becomes reality? In the morning I walked through thick humid air to the newspaper office, a ceiling-fanned bunker on the banks of a pleasure-boated river.

The features editor, Robin Doussard, introduced me to her assistant, Lisa Shroder. Both were bright, pleasant young women. When they concluded the interview, they asked if I had any questions for them. I was so ignorant of the running of a department that I didn't know to ask about budgets or special sections or editorial-to-advertising ratios; I asked them what magazines they read. I should have inquired, since I was going so off script, how many people in the newsroom had applied for the job. Travel editor was one of the most coveted positions at a newspaper as well as being the only one, among the specialty sections—Food, Fashion, Home & Garden—regarded as requiring no expertise. And, because of this, the job was traditionally given to someone as a reward for years of service writing news or business stories. I had expertise but, clearly from my question, no experience. They sent me out for a medical exam.

At the end of the day, Robin drove me to the airport. Dropping me off, she got out of her car and, across the roof, implored me to call her if I had any questions. Had it been a date, I would have taken this to mean she wanted to see me again.

As it turned out, the personnel office had a question. Robin called when I got back to Providence and said they wanted to see medical records of my eye.

The first Saturday in June, Hania and I boarded a chartered bus in Bala Cynwyd bound for the Polish consulate in Manhattan. While *my* life had been eventful of late, a lot had been happening in Poland as well, and this weekend the country was holding its first free elections since prewar independence.

Janusz, the organizer of the excursion, sat with his wife and three children, all of whom—with the exception of Janusz—wore T-shirts that read "*Filadelfia głosuje na Solidarność*" (Philadelphia votes for Solidarity). Many of the passengers worked at Penn—Janusz was an economist there—or the Wistar Institute. I'll miss this, I thought, if we move to Florida.

Solidarity's voting information sheets were passed around, along with copies of Timothy Garton Ash's article from the current issue of the *New York Review of Books*, "The End of Communism in Poland and Hungary." I got talking to an older man, a doctor by training, who told me he had come to the States for a year and stayed fifteen. "I still haven't acclimated," he said. "I don't have the necessary practical sense. I am a romantic." He was skeptical about elections but explained, "If I myself didn't vote, if I hadn't come today, I would have felt badly inside." He clutched his shirt in front of his heart and crumpled it in his fist.

A few minutes later the bus slowed abruptly, and people in the front shouted, "*Krowa!*" Looking out the window, I saw on the grass of the median a black-and-white cow. She was not grazing but lurching awkwardly backward in terror at the traffic. I had never seen a cow on the New Jersey Turnpike, and it seemed somehow appropriate that I was witnessing the spectacle while traveling on a bus full of Poles going to vote for the first time in parliamentary elections.

The bus rolled through the Lincoln Tunnel and headed to Madison Avenue and East 37th Street. A crowd of people milled outside the Beaux Arts building, while a constant flow moved through the open doors of the main entrance.

Hania and I made our way inside and climbed the red-carpeted, circular staircase. I had never been in this part of the consulate when applying for visas, and it was difficult to shed my usual feeling of angst. Though now the situation was deliciously reversed; rather than government officials holding our fate in their hands, Hania—and her compatriots—held the fate of Poland in theirs.

Upstairs, in an elegant reception room, Hania made her way to the M–Z stand and picked up her ballots. Then she waited for a booth to become free. A few less patient citizens filled out their ballots atop the closed bar.

When she emerged from the booth, Hania led me into the adjoining room. A portly official stood sweating profusely; it was impossible to tell if this was from the June heat or the new experience of having to be accommodating. Here people were using the top of a piano as a voting surface.

We returned to the main hall to observe the scene: women in summer dresses, grandmothers who looked as if they'd come straight from their Silesian villages, a large number of men with mustaches. Their light-hued summer clothing seemed at odds with their hard, somber, rough-hewn countenances. Shirts were worn with the flat collar open at the neck, and socks and sandals prevailed on feet. Despite the elegant surroundings, no one looked dazzled or out of place. In fact, there was an endearing casualness to the crowds that made me think of a Sempé illustration.

Outside, we lingered some more. Everyone around us was speaking Polish.

"I was an electrician in Poland," a man from Rzeszów told me, "but now I'm an auto mechanic. See?" And he held out a knobby, grease-stained hand unaccustomed to the environs of Madison Avenue.

There was something about the scene—the soft summer air; the labored faces; the pensive, not quite festive atmosphere—that seemed familiar. And then I remembered the pope's mass in Warsaw—ten years earlier, almost to the day. There as well, on a much larger scale, had been this same subdued, hopeful gathering, this coming together of national aspirations. And it was fitting that election day at the consulate should echo the pope's return to Poland because it was, indisputably, a direct descendant of it.

The following day we went to see Bob Spanier, my former editor at the *Observer*. I had given his name to Robin as a reference, and he asked if I had heard anything about the job. I told him they were worried about my bad eye, thinking it would be a handicap for a travel writer.

"But you describe things with your brain," the editor scoffed, "not your eye."

"Personnel says you're blind in one eye but you're such a good writer we want to hire you anyway."

It was Robin, calling from the *News & Sun-Sentinel* newsroom. She asked me to start at the end of August.

I put the receiver down, stood up, and walked around my cubicle in circles of bliss. Then I called Hania who, on hearing my voice, immediately assumed I hadn't gotten the job. My placid tone was the result of my guilt at uprooting her again. Philadelphia had grown on her, and she had become very close to Agnieszka and Elżbieta. (The rule about the difficulty of making friends after thirty bends when the friends are compatriots living abroad.) Later she told me I needn't worry; she was like her mother, who, after losing her family home as a child, never cared for material things again. Similarly for Hania: After the trauma of leaving Poland, any new upheavals would be insignificant. Never again would she feel emotionally tied to a place.

I walked over and told Phil, who appeared surprised and a bit puzzled; there was nothing admirable for him about travel writing—even though he knew it was my passion—*or* Florida, perhaps the one state he, like many Americans, found deserving of ridicule.

My last day at the paper, appropriately, was July 14th. The liberal columnist invited me to a farewell lunch at the Biltmore Hotel. When the check arrived, he suggested we split it.

A moving company loaded my belongings. Then I got in our new car—a blue Honda hatchback, with a stick shift—and drove out of Providence. The rare, twin joy—of leaving something endured for something desired—was similar to what I had felt on my departure from Greece, though this time it was professional as well as personal.

On my way to Philadelphia I stopped in Montclair to see Mike and Beth. Over glasses of iced tea, I shared with them my dream of producing a literate travel section, one with engaging stories as well as information.

I told Mike I'd love for him to write for it. They looked intrigued, though a part of me wondered if they were as dubious as they had been when I had told them I was just going to ignore computers.

In Philadelphia, our second household was put in the van. We drove to Elżbieta's house in Wynnewood, where Agnieszka and Krzyś Hariasz's wife Maria had come to say goodbye. At one point the women filled the front porch like a gathering of sisters—all of them short, with short, dark hair—and, watching from the street, I felt sadness for its dissolution and personal responsibility for being the cause. Then Hania and I got in the car and headed for I-95.

Three days later we arrived in Fort Lauderdale. I pulled into the numbered space in front of our new apartment building, and Hania reached over for a resolute hug.

Three months later the Berlin Wall came down. I thought of friends and family in Poland and of how, almost in tandem, my life and the world had taken turns for the better.

ACKNOWLEDGMENTS

Trenton: Sally Lane & Sam Graff, Dan Laskin & Jane Cowles, Mike & Beth Norman.

Greece: Christos & Linda Kontovounissios.

France: Dany Mall (Kutzenhausen) and Zbyszek Kossowski (Paris).

Warsaw: Jolanta Kuczma, Peter Meinke, Elżbieta & Jan Grzebalski, Gyuri Szőnyi, Anna & Jan Popiel, Kasia & Kubuś Michałek, Monika & Jurek Thieme, Andrzej Kaniewski, Zosia Puchalska Souche, and Monika Regulska.

Philadelphia: Agnieszka Baumritter, Elżbieta Sachs, Maria & Chris Hariasz, Bob Spanier, Gloria Klaiman, Kristen Yiengst, and Barrie Van Dyck.

Providence: Phil & Grace Terzian.

Florida & beyond: Robin Doussard, Lisa Shroder, Greg Carannante, Dave & Nancy Wieczorek, Mark & Cecile Gauert, Don & Joanne Dickerson, Bob & Betsy Pickup, Charles & Claudine Campi, Ela & Marek Kowalski, John & Christine Dolen, Katia Breslawec & Guy Peterson, Dave & Jen Seminara, Diana Abu-Jaber, Pamela Petro, April Eberhardt, Ericka Hamburg, Joe Recchi, Mitchell Kaplan, Cristina Nosti, my brothers and sisters-in-law—Bill, Jim, Pat, and Joyce—and of course Hania, who has given me such a rich and satisfying life.

Special thanks to the editors who published excerpts of this book: Sari Botton ("Letters from Trenton," *Longreads*), Jonny Diamond ("The Cold War Love Story of a Would-Be Travel Writer/Almost-Spy," *Literary Hub*), Sudip Bose ("Polish Lessons," *The American Scholar*), and Melinda Lewis ("Out of Philadelphia," *The Smart Set*).

I am extremely grateful to Pico Iyer for writing such a gracious foreword, and to the team at Rowman & Littlefield for their dedication and patience, especially my editors Katelyn Turner, Ashley Dodge, and Jenna Dutton.

This book is in loving memory of my parents, Winifred and Howard, and my friends Bob Joffee, Krzysztof Siudyła, Jeanne Meinke, Andy Mew, Vernon Young, and David Beaty.